THE LAW OF UNDULATION

A Believer's Journey of Faith, Perseverance, and Maturity through Life

William Femi Awodele

Foreword by
Pastor James Fadele

Copyright © 2011 by William Femi Awodele

The Law of Undulation
A Believer's Journey of Faith, Perseverance and Maturity through Life
by William Femi Awodele

Printed in the United States of America

ISBN 9781612158150

All rights reserved solely by the author. The author guarantees all contents are original and do not infringe upon the legal rights of any other person or work. No part of this book may be reproduced in any form without the permission of the author. The views expressed in this book are not necessarily those of the publisher.

Unless otherwise indicated, Bible quotations are taken from NIV. Copyright © 2005 by Zondervan.

www.xulonpress.com

Dedicated

To

Clive Staples Lewis (1898 – 1963)

Foreword

If there is to be full gratification in life, a child of God needs to experience the spiritual seasons designed by the Creator.

The Law of Undulation, skillfully written by Femi Awodele, is a book that will equip any child of God to forge ahead through the journey called life and its sinusoidal (undulating) waves. If you are experiencing moments of lows in your life, you are not alone and your situation is not unique – just ask Joseph, David, or Paul. The Creator has ways and methods that are like no other to bring His children to maturity. His specially designed spiritual seasons are used to bring His people to a place of elevation and upliftment for His own glory.

Femi Awodele encourages the believer who is in doubt about God's abiding presence and power to rejoice because God uses the seasons of our lives, both spiritual and emotional, for our promotion. Be encouraged – no situation is outside the scope of God, as seen in scriptures. "For thou hast possessed my reins: thou hast covered me in my mother's womb" (Psalm 139:13).

God's word gives a glimpse into some of the wonderful things He has in store for the believer. He has promised us a land flowing with milk and honey, but it is also a land of mountains and valleys; both are needed, and you must go through both to mature (Deut. 11:10-12). Through the ups and downs of Christian life, you need endurance to make it through and come out a winner.

Remember that His eyes are always upon you. No matter what tomorrow holds, He has inscribed you in the palms of his hands (Isa. 49:15-16). He will never leave nor forsake you – that is the blessed assurance.

Read and reread this book, and you will be greatly blessed. I also encourage you to share copies with your friends and family.

Pastor James O. Fadele
Chairman, Board of Coordinators, the Redeemed Christian Church of God, North America

Introduction

The whole family (except our dog, "Precious") was to spend the Valentine of 2004 in London, England, a trip precipitated by an invitation to speak at a marriage conference in Kennington, London, England.

Ola (my wife) always buys huge novels to read when we travel long distances. This time she ordered one book from Christian Books Distribution (CBD). The book came a day before our trip and as soon as we boarded the plane at the Chicago O'Hare airport, she started reading and kept to herself for the most part of the trip (which was unusual). As soon as she was done, she told me I had to read the book.

I could not read the book while in England due to my engagements and sightseeing with the boys. As soon as we boarded the plane in Heathrow, she handed the book to me, really against my wish because I just wanted to sleep. As I opened the pages of *The Screwtape Letters* by C.S. Lewis, I became fascinated and actually finished it before we landed at Chicago O'Hare airport. The story was fascinating to me. I especially found a line of the book relevant to what Ola and I were going through at the time. The line was "the law of undulation."

The Screwtape Letters is about two demons. The senior demon, Screwtape, is a mentor to the junior demon, Wormwood, who has been assigned to a new believer. The two demons exchange letters as to how best to turn the believer back to his old ways. In one letter, the junior demon wrote how the believer had been cornered due to a "problem" he was going through. In his letter back to the junior demon, Screwtape scolded Wormwood about how a valley experience can be the best thing to happen to a believer and then used the phrase "the law of undulation" as a principle that God uses in a Christian's life.

This supposed principle described in the novel by C.S. Lewis disturbed me for the rest of the trip back to Omaha, and I could not wait to study the Bible and prove or disprove this principle. To my amazement, I found that not only is the principle true but it is also applicable even today in every Christian's life. The better we understand it, the more we'll enjoy this journey called life.

It is my prayer that this book will change the life of everyone who reads it, just as it touched the lives of church members and groups as I preached a sermon on it in 2004.

William Femi Awodele

Acknowledgement

This book came out of the ingenious gift of narrative communication that God placed in Mr. Clive (C.S.) Lewis, a man who converted to Christianity in a rather unusual way. His approach to Christianity is clearly related to his academic background, definitely different from my experience of being born into Christianity, so I thank heaven for the gift placed in Clive Lewis for the body of Christ.

I recently realized that my role in the body of Christ, whether speaking about relationships or any other topic, is to challenge the body of Christ to come back to strict Biblical doctrinal principles along with a life that strives for obedience, and to not be carried along by changing culture in the name of tolerance or of making church folks comfortable. This has not been an easy assignment, so I appreciate those who quietly encourage me when I think I'm crazy or think I'm not qualified for the assignment.

I appreciate the many lives worldwide that have allowed God and His word to come into their lives and homes through the articles either on Nigeriaworld.com or in newspaper and magazine outlets. My prayer is that the Lamb that was slain would receive the reward of His suffering in your lives and through your obedience and integrity in your home.

To my Abike Peperempe, you are greatly appreciated as you continue to challenge my walk with God. To

Ibukunoluwa, you are indeed what your name says – you are a blessing to me and your mom; we are very proud of you and your relationship with God. To Fiyinfoluwa, your bright view of life is contagious and I pray you never lose that; many will one day glorify God in heaven because of your great attitude toward life. To Ebunoluwa, you have brought a different perspective of being the hand and leg of Christ to other humans – I can't wait for the day your brothers, Mom, and I attend your graduation (as a Veterinarian).

To our parents, Dad and Mom Awodele, and Dad and Mom Songonuga, it is my prayer that God will continue to strengthen all of you. May your later years bring more joy than your former years. To siblings, cousins, nephews and nieces, this includes you TJ thanks for being family.

To my friend and brother in the Lord, Pastor James Fadele for writing the brilliant foreword and to Joanne Chapuran, your patience with me through the editing process is greatly appreciated.

William Femi Awodele

TABLE OF CONTENTS

Introduction ix

Acknowledgement xi

Chapter One .... Why Undulation? 15

Chapter Two .... Recognizing a valley 27

Chapter Three .... How we get into a valley 35

Chapter Four .... The enemy's strategy for a
                      believer in the valley 51

Chapter Five .... What to do in the valley 67

Chapter Six .... How God views your valley 84

Chapter One

Why Undulation?

Undulation is defined as a wave-like motion, a gentle rising and falling in the manner of waves, forming a valley- and mountain-like shape (according to my computer's dictionary). In physics, undulation is defined as a motion to and fro, up and down, in a fluid or elastic medium, propagated progressively among its particles without the displacement of the medium.

I like the physics definition because it is based on laboratory experiments and it accurately describes what I want to establish in this book. There are particles like wind that cause the up and down movements of a fluid or elastic object; however, in all the movements the medium itself is not displaced. Also, this movement of the fluid eventually benefits the medium. For instance, the constant movement of the ocean brings about the cleaning needed; stagnant water eventually becomes stinky, dirty, and attracts many disease-causing insects.

It is my belief that in Christians' lives, God allows these wave-like situations, often called valley and mountaintop experiences. King David used this analogy in Psalm 23:4: *"Even though I walk through the valley of the shadow of death, I will fear no evil,"* and in Psalm 36:6: *"Your righteousness is like the mighty mountains."* Sometimes mountains also represent an obstacle, such as in Zechariah 4:7.

Why would a good God allow valley experiences in the lives of His children? The answer is found all over the Bible, but let's consider two scriptures. Deuteronomy 8:2 says, *"Remember how the Lord your God led you all the way in the desert these forty years, to humble you and to test you in order to know what was in your heart, whether or not you would keep his commands."* James 1:2-4 says, *"Consider it pure joy, my brothers, whenever you face trials of many kinds, because you know that the testing of your faith develops perseverance. Perseverance must finish its work so that you may be mature and complete, not lacking anything."* According to Apostle James, a member of the Jerusalem council and the brother of Jesus, trials are supposed to do three things in a believer's life:

1. Develop our faith
2. Develop perseverance (or patience)
3. Mature a believer (spiritually, emotionally, and in other ways)

Let us consider some Bible individuals that show these three purposes for undulation in a Christian's life.

Saul, Solomon, Joseph, and David are all important people in the history of Israel. Two of them ascended to the throne with no real tests in life. In fact, the stories we know about them before they became VIPs do not make good resumes. One was a quitter while the other was mama's boy. The other two literally went through the valley of the shadow of death which adequately prepared them for their destination, and people today have written thousands of books about both of their leadership styles.

Saul, son of Kish the Benjamite as King of Israel

After getting to the Promised Land, the Israelites were ruled by judges and prophets at first. To keep up with the Joneses around them, they asked God to give them a king. Even after warnings about what the king would do, they still insisted on getting one. God told Samuel to anoint Saul from the house of Benjamin as the king.

Our first introduction to Saul was not a good one in my opinion. His dad, Kish, told him to go look for some lost donkeys with one of the household servants. He looked for a while, but then told the servant that they should give up looking for the donkeys since they could not be found (1 Samuel 9:5). This sounds to me like a five-year-old who is told to take something to his room while he's watching the Cartoon Network. Of course, he won't. The task is not as important to the child as to the parent. Children are in a hurry to get back to selfish fun. This introduction of Saul shows that he has not been tested in life; he still gives up on life when it becomes too difficult (which is what he exhibited many times as the king of Israel).

This pattern of childish behavior or lack or maturity was seen throughout his life. Saul never had a real test that prepared him for the position in which he found himself. In 1 Samuel 13, we read about his lack of maturity. As the king of Israel, he camped at Gilgal with his army and was waiting for Samuel the priest to come and offer a burnt offering to the Lord before they went into battle. He waited for what seemed like an eternity to him (actually seven days) because his soldiers were afraid. He went ahead and presented the burnt offering himself, something reserved for the priest only. In 1 Samuel 15, we read of God instructing Saul the king of Israel to go into battle and destroy everything that belonged to the Amalekites (the Amalekites' cup of sin was full and God was using this battle as a judgment). Instead of obeying God and destroying everything in battle, Saul preserved the good animals and Agag the Amalekite King. When Prophet Samuel confronted him about his disobedience to God's instruction, Saul gave a "childish" excuse, saying that his soldiers had brought the animals back and that he wanted to use the animals as sacrifices to the Lord. After this, Samuel told Saul that he had been rejected as king by God. He acted childish yet again; he begged the Prophet Samuel to appear with him in front of the elders and the people so that it would look like he was still in charge of the kingdom.

As the king of the land, King Saul was someone whose word literally became law without congressional or parliamentary debates. To show you how immature King Saul was, he was jealous of a teenager who helped his government by killing the giant that was hurling insults at his (Saul's) army.

Solomon king of Israel, son of David

Without a doubt Solomon was the favorite of his father King David; his name even means "the Lord loves him." He must have been protected from the likes of Absalom, Amnon, and Adonijah, his older brothers who were dangerous and had been to wars with their dad.

One thing was clear in how David's sons behaved – as much as he loved the Lord he never really passed down that heritage of loving the Lord with all of his heart to his children. Absalom wanted to kill his father and get the throne; Amnon raped his half sister (Tamar) and was eventually killed by his half brother Absalom. Another son, Adonijah, plotted to overthrow his dad while David was on his sick bed in his old age; this same son had the gall (obviously a lack of respect for his father) to ask to marry Abishag, his father's wife in his old age.

Solomon came from this dysfunctional home, but his dysfunction was more in the area of not being exposed to real life. Instead he was shielded, more like a last-born child who escapes disciplining by parents as they grow older. His father loved him so much and on his deathbed told Solomon to make serving Jehovah his priority. Serving Jehovah lasted a while for Solomon – he did build the temple that David had wanted to build (and had building materials ready) and he was blessed with wisdom and wealth – but from then on, he made a 180 degree change by marrying lots of women and serving the women's foreign gods. God was so angry at Solomon's serving other gods because of his many wives that in 1 Kings 11:11, He said, *"Since this is your attitude and you have not kept my covenant and my decrees, which I commanded you, I will most certainly tear the kingdom away from you and give it to your subordinate."* But because of God's love for David and His covenant about Jerusalem, a kingdom (Judah) was kept for the descendants of David.

For someone who started on top, Solomon did not finish well at all. In fact, the Bible says that God raised up adversaries, allowing his enemies to cause unrest in the kingdom.

From studying the lives of Saul and Solomon, we see that both men attained leadership positions (definitely because of God's love), but their lack of faith, lack of patience, and immaturity in spiritual things eventually made them poor examples of leadership. These things they lacked can only be gained through tough life experiences.

Now let's study two men who attained leadership positions but went through many tough situations (valleys) in life which developed their faith, patience, and matured them before they became leaders of men.

Joseph, son of Jacob (Israel), Prime Minister of Egypt

This 17-year-old kid, loved by his dad, was not spoiled but was put to work. According to the Bible he was a shepherd from an early age, tending to the family animals. His step-brothers did not like him, however, because he was Dad's favorite (although this was something he did not ask for; it was out of his control). Jacob probably loved Joseph because he was a child who demonstrated Jacob's physical abilities in the bedroom at his old age (something to be proud of). He was also a child from Jacob's favorite wife.

To complicate matters for Joseph, he had two dreams that revealed that his family (especially the brothers who were already jealous of him because of a lousy ornamented coat) would one day bow to him in worship. The thought of bowing to Joseph was so chilling to his brothers that they plotted to kill him and would have, but for the Holy Spirit moving through Reuben (one of his brothers). They eventually spared his life, but he was sold into slavery, and they told their dad that Joseph had died in the wild, perhaps killed by a wild animal (which was their original plan).

The Midianite merchants who bought Joseph from his brothers sold him as a slave to Potiphar, one of the most powerful people in Egypt and an official of Pharaoh's monarchical administration. Joseph did not allow the bitterness of what his brothers did to make him go into depression (like a lot of us do these days); his faith in his Jehovah kept him going. He was a young, handsome man in charge of Potiphar's house, and his master's wife soon noticed his build, intelligence, and charisma (which he must have had to be a leader) and wanted a part of that. She tried seducing Joseph (to have sex with her), but because of his faith and belief in his God, Joseph refused.

As a young man, I was raised in church but rebelled from it during my high school and college years. As an adolescent with testosterone driving my body, I remember having a science teacher that was really beautiful. A lot of us boys wanted to be around her. If she had asked me out then, I doubt if I would have refused. The above is what is happening in our society now with high school female teachers seducing their male students, some even having children by these "boys."

Potiphar's wife probably thought it would be an honor for Joseph to be invited to bed by his master's wife. Well, she was wrong, and the shame of rejection made her frame Joseph for attempted rape, which landed him in prison. Again we see Joseph's faith coming into play, as he did not hold a grudge against Potiphar or his wife. He kept trusting God, and he soon became the chaplain's assistant in prison. (As someone who has been blessed to be in prison ministry for 18 years, I know the liberties that come with being the leader inside the prison or the chaplain's assistant in jail – they get privileges from the officers and get to deal directly with volunteers from the outside. They also get to distribute items brought inside by volunteers, such as soap and such which are hot commodities, especially in African prisons where the governments don't care about the prisoners' welfare.)

Joseph's prison was a club fed (prison for ex-government officials) and he soon became the popular guy because of his administrative abilities and God's hand on his life. He interpreted the King's ex-cupbearer's and baker's dreams to them. He said that the baker would be killed by the king and the cupbearer pardoned. He then asked the cupbearer to talk to the king upon his release because he had not committed the crime for which he had been jailed. In life, when people climb the ladder of success, they easily forget what brought them to their present position and start to behave pridefully or stupidly. The cupbearer, of course, forgot all about Joseph once he was pardoned and got back to the service of Pharaoh. It took two years and the king having a dream before the cupbearer remembered Joseph. Joseph learned patience. (Two years in jail or prison ain't no joke, club fed or not.)

With the help of God, Joseph interpreted the king's dream and eventually became the Prime Minister of the most powerful country in the world. The Bible records that Pharaoh trusted Joseph with the whole kingdom, and that Joseph managed it well.

When Joseph was in a position to avenge his brothers' selling him, he forgave them. Joseph was mature enough to know that with God, *"all things work for the good of those who love God and are called according to His purpose"* (Romans 8:28). Having learned patience while in prison and keeping his faith in Jehovah throughout the whole ordeal, as Prime Minister, Joseph had matured enough to know the heart of God about forgiving his brothers when they came to Egypt to buy food.

David son of Jesse, King of Israel

This was another teenage shepherd who had fought bears and lions while protecting his family's animals.

When God told Prophet Samuel to go to Jesse's house to choose a king, no one reckoned that little David would be God's choice. His dad and Prophet Samuel were actually looking at his big brothers Abinadab, Shammah, and the seven others because of their physical appearance.

David had been put to the test many times in his life, even as a teenager. Being a shepherd is tough enough, but fighting off lions and bears to protect the sheep and goats is another thing. He had faith in his God that no matter how mean the animals looked, he would defeat them. This same faith worked for him when he visited his brothers at the warfront.

The army of Israel was being abused on a daily basis by a giant named Goliath, who led the Philistine army. Not one in Israel's army had the faith to fight this guy because they'd never had to exercise their faith. Even their King, Saul, was used to looking for any way out of situations. On getting to the army of Israel's camp, David saw no fighting; what he saw instead was a bunch of cowards letting an uncircumcised Philistine blaspheme the name of the Lord God of Israel. David went to King Saul and said, *"Your servant has killed both the lion and bear; this uncircumcised Philistine will be like one of them, because he has defied the armies of the living God. The Lord who delivered me from the paw of the lion and the paw of the bear will deliver me from the hand of this Philistine"* (1 Samuel 17:36-37). There is no doubt that David's experience in fighting the wild animals gave him the boldness to want to fight this giant. He also recognized that the strength given to him by God was still available to defeat the giant.

Upon David's defeat of the giant, King Saul became jealous because David was now the "Hero." It was as if every neighborhood had a street named after him – David Boulevard, David Avenue – and songs were composed in David's honor, especially after he won even more victories over the Philistines. The King started plotting ways to kill David. He tried himself twice with an arrow, he tried using his own daughters, his commanders, and even his own son Jonathan. David escaped the plots to kill him and was on the run from Saul for about a decade, with Saul in hot pursuit.

Twice he had the opportunity to kill Saul, and on both occasions he spared his life, again displaying the maturity that comes only from fearing the Lord (Psalms 111:10). David learned patience while running away from Saul. David had been anointed king in his teens, yet he refused to kill the current king when he had the opportunity (unlike Macbeth in Shakespeare's story). He learned to wait on the Lord. He learned to trust God and to seek God's face for every decision he had to make. He had rag-tag army of 600 and realized that his victories in wars were definitely because of God, not the strength or training of his men. He wrote many poems while on the run and we read a lot of them today in the book of Psalms.

Many of the lessons he learned while running served him well as a King. As a King, he inquired of God before he did anything; the only time he didn't, he paid dearly for it (2 Samuel 24:1-17). His trust in the Lord must have been built up by the many near-death experiences he went through – braving the wild animals and Saul's attempts on his life, living and finding favor among the Philistines (his arch enemies after he had paid his bride price for Michal, daughter of King Saul, by killing 200 Philistines). He had a relationship with the Lord like no one else because he had to. God had been faithful to him many times. David knew God and what God wanted.

From studying the lives of these four men, we notice that the two (Joseph and David) who went through life's school of hard knocks before arriving at their destinations ended well; they had good skills and were loved by God, even with character flaws like David had. God was so pleased with these two guys that they got more blessings in life than their counterparts. Joseph's sons Ephraim and Manasseh both got portions of the Promised Land (a double portion), while Joseph's brothers had only one each. Because of David's love for God, He (God) made a covenant with him that his throne would last forever, a covenant fulfilled by Christ being born in the lineage of David. The other guys (Saul and Solomon) pretty much had the kingdom dropped into their laps and never struggled or went through anything to appreciate what they had. They both ended badly, with God rebuking and rejecting them.

When we go through trials in life, they are designed to strengthen us for whatever lies ahead of us. When Joseph was serving about a two year sentence in prison, he was not thinking about learning patience, even though he learned it. David had to inquire of God for every battle he fought while in exile because he had 600 untrained guys, so he got used to leaning on the Lord.

The Law of Undulation

The writer of Hebrews emphasized the importance of faith development and how without faith it is impossible to please God (Hebrews 11:6). Hebrews shows the fathers of faith persevering, waiting for the blessings, and maturing enough in their walk with God to be commended as ones who pleased God.

The above principles of faith through trials can be applied in our days. For example, attaining a position because of family lineage and not because of skill or traits often brings disaster; the person is not tested and may not be qualified. In the same way, sudden wealth does not usually last.

Let's consider the following lottery winners and what happened to them because they were not ready for their sudden wealth (Goodstein 2006).

1. Suzanne Mullins won $4.2 million in 1993. Today she is in debt, and she blames the illness of her uninsured son-in-law for her financial problems.

2. Willie Hurt of Michigan won $3.1 million in 1989; two years later he was broke and was charged with murder. Much of his money was spent on divorce and crack cocaine.

3. Charles Riddle of Michigan won $1 million in 1975; he later got divorced, faced several lawsuits and was indicted for selling cocaine.

4. Janite Lee of Missouri won $18 million in 1993; he was so generous that eight years after his winnings, he filed bankruptcy with $700 left in two bank accounts.

5. William Post won $16.2 million in 1988; he now lives on social security, $450 a month, and food stamps.

Another example is sports superstars who make huge amounts of money and waste it on frivolous things. Consider a basketball star who jumped from high school to the pros and many years later has had many kids from different mothers and plenty of lawsuits; he now plays for a second rate pro team and is only a shadow of his former self. Many sports, music, and movie stars have filed for bankruptcy because they were not prepared in all aspects of life for the sudden riches they got.

The richest man in the world (2005) lives in a small city in the Midwest, he drives a modest car and his house is not the priciest in the city; in fact, it's nowhere close in size or flashy exterior to the golf course neighborhoods that a lot of wannabes live in. The second richest man in the world is on record that he will leave most of his money to charities and little to his kids because of the potential damage to his kids' lives.

Chapter Two

Recognizing a Valley

Valleys come in different shapes and sizes. They could be geographical, emotional, physical, financial, or involve other areas of life.

Moses

Moses, undoubtedly a central figure in the Bible, went through a life journey of valleys and mountaintops. The tough valleys he went through prepared him for his purpose in life, which was to be God's instrument in librating the Israelites from slavery in Egypt.

In the book of Exodus we see that the valley in Moses' life started way before he was born because of his destiny. After the death of Joseph, who was the Prime Minister of Egypt, a new Pharaoh who did not know Joseph (and probably did not care) came to power and enslaved the Israelites. This period was an **emotional and physical valley** in the lives of the Israelites living in Egypt. Prior to this king, life was good for the Israelites; they grew in number, food was plentiful, kids went to school, and things were great for a period of time. Then trouble started. They were forced into slavery because the new king was afraid of their numbers (a perceived threat). The slave masters dealt with them ruthlessly, yet they continued to grow in number.

Their growth and success so marveled Pharaoh that he ordered Egyptian midwives to kill Israelite children during birth. The midwives, however, could not follow the king's order because they feared God (Exodus 1:17). The principle here will be discussed in detail later, but while the Israelites were in this valley, God still had his hand on them. The king then ordered his soldiers to kill the male children. It was during this valley in the lives of the Israelites that Moses was born. His mom, fearing for his life, hid him at home for three months before finally deciding to trust God with his life. She put little Moses in a basket and set him in the water at the mercy of the wind. With God choreographing the whole event, the wind took Moses' basket to the Pharaoh's palace, where he was taken care of for forty years of his life. He went to the best schools and enjoyed other luxuries that came with being in the royal family, basically enjoying a mountaintop experience for forty years. Somehow Moses realized he was adopted into the palace and that he was an Israelite. One day while watching the Israelites work in the field, he saw an Egyptian supervisor beat a slave. Thinking there was no one watching, he killed the Egyptian. The next day what he did was reported to his face. Afraid for his life, he fled Egypt.

Moses fled Egypt and went to Midian as a shepherd, creating a **geographical valley** in his life that lasted another forty years. In retrospect, one can argue this was part of God's design in his life (and I totally agree), but while Moses was living in Midian in tents, moving animals from one place to another for pasture, it was definitely not fun for him. Imagine living forty years of one's life in the Astoria Waldorf Hotel in New York City, then for some strange reason living the next forty years in a Motel 6 in a broken down inner city. Imagine someone living in an affluent suburban golf neighborhood for forty years, then living in a trailer park in the other end of town for the next forty. It must not have been fun for Moses. I was born and raised in Lagos, Nigeria and I loved every minute of my growing up; I never thought I was missing anything. At age 26, I moved to America and have lived in American cities for over seventeen years now. Whenever I go back to Nigeria for one or two weeks, the lack of running water, good roads, and adequate communication really get on my nerves; I start to miss America.

The forty years as a shepherd in Midian taught Moses life lessons that he would never have learned had he stayed on at Pharaoh's palace; he would have thought he was invincible, like most politicians think they are today. Living in an uncomfortable environment that was different from what he was used to humbled him and drew him close to God.

Abraham (Abram)

Terah, Abram's dad, left Ur of the Chaldeans for Canaan with his son, his son's wife Sarah, and his grandson, Lot. They, however, settled in Haran, where they became very successful, owning many animals and slaves (Genesis 12:5). Terah died and left all the possessions to Abram, his son. God then showed up on the scene, telling Abram to leave Haran (his comfort zone) to a land he would be shown (a **geographical valley** for Abram), and God promised to make him the father of many nations and give him the land of Canaan as his inheritance. While God gave Canaan to Abraham's descendants (Joshua 14), Abraham himself lived in many different locations, moving from place to place, as he was directed by God.

The promise of a child was definitely an **emotional valley** for Abraham. God promised to make him the father of many nations at age 75. Along the way he had Ishmael (son of Hagai, Sarah's Egyptian maid). Finally at age 100 (when Sarah was 90), the promised child came and was named "laughter" (Isaac). As the promised child grew and Abraham started getting comfortable, he got yet another visit from God. This time he was to take the promised child and sacrifice him (Genesis 22:1-2). As a father and a believer in the Lord, I can empathize with Abraham that he had the toughest decision a father could have in this world. However, Abraham had related with God so many times that his faith in God was unshakable. First, he had left Haran, secondly, his nephew was saved from the destruction of Sodom and Gomorrah, and then his wife Sarah (way past menopause at 90 years of age) had a son. He had also learned perseverance as he wandered in the desert, moving from one place to another.

One thing that fascinates me about Abraham's life is that he never had a financial valley; he understood the principle of God's blessing, which involves obedience and giving. He gave to Melchizedek (Genesis 14:17-20) when no one ask him or forced him to do so.

Samson

Samson is another guy in the Bible (Judges 16), who enjoyed mountaintop experiences all his life until he told Delilah the secret of his strength and was thrown into a Philistine jail. Samson was a powerful guy, feared by enemies and respected by his own people. When he was arrested, his eyeballs were gouged out; he was bound with bronze shackles and was set to grinding in prison. This was definitely a **physical, financial, geographical and emotional** valley for him. (Just imagine Saddam Hussein's picture in his palaces followed by images of his capture by American soldiers in a hole and his shabby appearance defending himself in a courtroom.)

David

David was one guy who had a lot of different valley experiences in his life. After defeating Goliath and winning several wars for Israel, King Saul was jealous of David, who had to run away for his life. He lived in caves and in other tough conditions (a **geographical valley**), away from his wife, Michal, and other family members (an **emotional valley**). While on the run with his army, they had financial problems (a **financial valley**) that led him to dealing with Nabal the Calebite and eventually to his marriage to Abigail.

As the king of Israel, he was again put on the run (a **geographical valley**) by his own son Absalom (2 Samuel 15:13, 16, 17). I believe also that the behaviors of his children, raping and killing each other (2 Samuel 13:31), must have brought great discomfort to him (an **emotional valley**). David's affair with Bathsheba, the death of her husband Uriah, and the Lord's rebuke brought another **emotional valley** to the life of David (2 Samuel 12:13-17).

David was a master at getting into valleys and back on mountaintops (see chapter 3).

Daniel, Shadrach, Meshach, and Abednego

King Nebuchadnezzar had just defeated Judah (a valley in the life of the Israelites as a nation) when the king instructed his men to identify smart and handsome Israelites to be trained in Babylonian language so they could serve under his administration. Four such men were mentioned in the Bible—Daniel (Belteshazzar), Hananiah (Shadrach), Mishel (Meshach) and Azariah (Abednego).

This period of captivity was a low point in the life of these men. To add insult to injury, King Nebuchadnezzar made an image of gold of himself and asked everyone in his kingdom to worship the image. According to the laws of

Moses, Israelites are never to bow to another God, so to bow to this image was a violation of their God's commandment. For refusing to bow to the golden image, Shadrach, Meschach, and Abednego sunk to yet another level in the valley. They were threatened with being thrown into a fiery furnace, to which they replied, *"If we are thrown into the blazing furnace, the God we serve is able to save us from it, and he will rescue us from your hand, O king. But even if he does not, we want you to know, O king, that we will not serve your gods or worship the image of gold you have set up"* (Daniel 3:17-18). They were thrown into the fire, but God rescued them while in the furnace. Daniel encountered his own test in Babylon when other leaders plotted to have him killed because they were jealous of him. Their plot required that he be thrown into the Lion's den. According to Daniel 6, the Lord shut the mouth of the lion when Daniel was thrown into the pit.

Peter

Denying Jesus three times after Jesus' capture was an emotional valley in the life of Peter, (Mathew 26:69-75). The Bible said that after the rooster crowed, Peter went outside and wept bitterly.

In our lives today, perhaps as you read this book, each one of us is going through a valley in one or more areas of life.

There are people going through the loss of a child. I know a family who recently lost their teenage son; everything was going fine up to that point, then for weeks the family kept asking questions about why God allowed such a thing to happen to their family. Losing family members is not fun, even if they die at an old age. My maternal grandma died at 69 years old, so she wasn't really too old. Her death was especially tough for me because it happened during my rebellious years. Being a Godly woman, she would call me and pray with me. We had a conversation before her death that we

The Law of Undulation

did not finish before I went back to school in another state. I still wonder what her strategy was to correct her rebellious grandchild. (It probably worked; I'm in ministry today).

Some families have never experienced the death of a child, but the marriage has never known peace. I know of marriages that God has blessed with finances, power, and worldly influence, but there is no peace in the relationship, and their kids are rebellious and on drugs. As a youth prison clergy volunteer some years back, I met many young adults from good homes spending time in a detention center because of peer pressure or the influence of drugs in their lives. A wayward child is a valley in the life of the parents.

Any form of sickness or illness is a valley. Bouts of depression, cancer, an autistic child, high blood pressure, diabetes, or any other health related issues can bring great discomfort.

Major catastrophic events are also major valleys that can last a long time. Like Job, many families have gone through events that seem so deep they think that they'll never crawl out. The author of *His Beauty for My Ashes*, Rev. Tai Ikomi, wrote about losing her husband and three kids in an accident caused by a drunk driver; the car burst into flames and she watched her family burn to death with the kids screaming. In August 2003, Robert Rogers and his family were traveling on a Kansas highway when a flash flood engulfed their car; he lost his wife and four children in that accident. These stories tell of scars from a very deep valley. (We'll talk about healing from valley scars in later chapters.)

Living in an industrialized country is definitely better than a third world country, yet it can still be a valley in the lives of many immigrants. Immigrants come to America in search of a better life, but some actually end up with a worse lifestyle. I remember meeting a gentleman who was a high court judge in one of the island nations not far from the American mainland; when I met him he was looking to either train to be a nursing assistant or drive a cab. He had not thought that going for a major retraining was in his future. I personally wonder if such immigrants are not better off in their country of origin.

Our lives involve different roles – father, mother, child, husband or wife, employer or employee, a layperson or ordained reverend in ministry. Many times one role, such as employer, is in a valley while another, such as husband, is seemingly fine. There are people who are financially wealthy but are dealing with health issues that make their wealth useless. There are people with the peace of Christ as their children are doing well, but they are struggling financially. When Moses discovered he was a Jew while still living in the palace, he had turmoil in his spirit even though he still had wealth and influence.

Valleys are those times in our lives when at least some aspects of our lives are upside down. It can seem like the wall is closing in on us and nothing seems to work. It could be caused by any number of things, such as health problems, financial problems, extended family issues, the children, or a career-related issue.

Chapter Three

How We Get Into a Valley

We have examined how God uses the valleys we go through in life as life lessons, and we've also discovered that those valleys actually shape us. Chapter two helped us identify those valleys in the areas of our life.

In this chapter we will examine the ways we get into a valley. In my study of the Bible I have found three reasons we fall into a valley:

1. Time for spiritual promotion and emotional growth – James 1:2-4: *"Consider it pure joy, my brothers, whenever you face trials of many kinds, because you know that the testing of your faith develops perseverance. Perseverance must finish its work so that you may be mature and complete, not lacking anything."*

2. God's discipline – Hebrews 12:7: *"Endure hardship as discipline; God is treating you as sons. For what son is not disciplined by His father?"*

3. Sin – Roman 6:11-12, 23: *"In the same way, count yourselves dead to sin but alive to God in Christ*

> *Jesus. Therefore do not let sin reign in your mortal body so that you obey its evil desire...For the wages of sin is death."* James 1:15: *"Then, after desire has conceived, it gives birth to sin; and sin when it is full-grown, gives birth to death."*

Let's look more closely at these three ways of falling into a valley.

1. Time for spiritual promotion and emotional growth

The enemy of our soul is the devil and the scripture is clear about the devil's intent. 1 Peter 5:8 says, *"Be self-controlled and alert. Your enemy the devil prowls around like a lion looking for someone to devour."* We see many times in the Bible when the enemy seeks to destroy God's people. King Balak seeks to destroy the Israelites (Numbers 22), Haman plots to destroy the Jews (Esther 3), and Jesus said it best in Luke 22:31: *"Simon, Simon, Satan has asked to sift you as wheat, but I have prayed for you, Simon, that your faith may not fail."* From this statement of Jesus to Peter and from other examples in the scriptures, we know that the devil continually prowls, looking for a loophole in our lives through sin, or asking for permission to test us like he did Jesus Christ (Luke 4:1-13).

Allowing the enemy to test us is not the same as God tempting or sending evil into our lives. The Bible clearly states that God is not the author of evil. James 1:13 puts it this way: *"When tempted, no one should say, 'God is tempting me,' for God cannot be tempted by evil, nor does he tempt anyone."*

As children of God we are protected by God's mighty hand and absolutely NO EVIL can come near us except that which is allowed. Let's consider this conversation between God and Satan about Job: *"Then the Lord said to Satan,*

*'Have you considered my servant Job – there is no one on earth like him; he is blameless and upright, a man who fears God and shuns evil.' 'Does Job fear God for nothing?' Satan replied, **'Have you not put a hedge around him and his household and everything he has**?'"* (Job 1:8-10a). Let's consider yet another covering of God. In Numbers 22, we read about King Balak of Moab who was afraid of the Israelites because they were camped near his country and he had heard about their God and His mighty power. Thinking he was wise, he chose not to attack militarily but to put a curse on the Israelites. He decided to consult with a diviner called Balaam. In chapters 23-24, we read that Balaam tried cursing the Israelites four times and all four times he blessed them and prophesied into their lives. King Balak was mad at him and Balaam tells King Balak in Numbers 24:12-13: *"Did I not tell the messengers you sent me, 'Even if Balak gave me his palace filled with silver and gold, **I could not do anything of my own accord, good or bad, to go beyond the command of the Lord** – and I must say only what the Lord says'?"*

From the story of Job and Balaam, we can conclude that the enemy realizes that he cannot touch a child of God unless he is permitted. From Balaam's statement we can note that he secretly wished God would allow him to curse the Israelites so he could benefit financially.

The reason God allows some tests in our lives is so that we can continue to grow in our faith, knowing that without faith it is impossible to please Him. Abraham's faith increased to the level that he believed his wife would still have the promised child when she was way past menopause, and later he was even ready to sacrifice his son. In all of these situations, faith was credited to him as righteousness.

Promotion and growth brought through faith always come at a price. I have not always being an academic person; I remember the promotion exam from my third year in high school to the fourth year in Nigeria. I had been a better-than-average student until my third year when I joined a gang and got into girls. My grades dropped dramatically and my parents, looking for incentives to get me interested again, promised me a wristwatch if I got promoted to the fourth year. In Nigeria, the fourth and fifth years of high school are very important because they determine if you will be in science, commercial, or arts classes, which eventually determine if you will be a doctor, accountant, or lawyer.

The promise of a wristwatch spurred me on, and I did some crash reading that kept me up for about a week. I got promoted to the next class, even to the science track. As a medical student I saw my wife study for USMLE exams and eventually the board certification exam. Becoming a board certified physician is challenging; try asking physicians about their experiences during medical school and residency. Getting a promotion either at school or at work comes with some hurdles.

Allow me to share the story of two people who are close to me. One is an uncle who was a professor of medicine for many years. When the time came for him to earn tenure as a professor, someone within the faculty blocked his bid; a bitter battle then ensued. Frustrated, he went back to residency as a family physician (after teaching anatomy for many years). Going back to practicing medicine rather than teaching medicine became God's plan after all, as the income of a practitioner is more than that of a teacher. It helped him put his kids through college without incurring any debts.

The second person is my wife. We moved to the American Midwest because we believe it is a good place to bring up our boys and it caters to her love of delivering babies. She got a job with a group of primary care physicians. Two years into working with this group, she told me that she was not practicing medicine the way she wanted to. In residency, she was awarded the Most Compassionate Resident because of the time she spent with patients and her down-to-earth explanation of medical terminology. In this group, she could not do that because of the emphasis on finances (bonuses, overhead costs, etc.). Two years later she was let go for improper documentation (a charge she was absolved of by the state investigator, who determined it was an error). While considering her next move, she came across a new concept in medicine – a small office, electronic documentation, and no staff (see AAFP articles on Solo Practice – Dr. Gordon Moore). Today her practice is thriving; she sees each patient for 30 - 45 minutes and gets to explain as much as she wants. Most of all she has the time for her family and her God, something she had been struggling with while in the group.

The writer of Hebrews, talking about the earthly ministry of Jesus Christ, said, *"Although he was a son, he learned obedience from what he suffered"* (Hebrews 5:8). Even Jesus the Christ on earth was not immune from learning through suffering. In Matthew 26:36-46, twice Jesus said, *"My father, if it is possible, may this cup be taken from me. Yet not as I will, but as you will."* He understood his mission in the world (John 3:16, Hebrews 8) and knew that to get there, there would be suffering. Philippians 2:5-10 says:

> *"Your attitude should be the same as that of Christ Jesus: who, being in very nature God, did not consider equality with God something to be grasped, but made himself nothing, taking the very nature of a servant, being made in human likeness. And being found in appearance as a man, he humbled himself and became obedient to death – even death on a cross! Therefore God exalted him to the highest place and gave him the name that is above every name, that at the name of Jesus every knee should bow, in heaven and on earth and under the earth, and every tongue confess that Jesus Christ is Lord, to the glory of God the Father."*

The Law of Undulation

Jesus Christ on earth went through many valleys, being rejected in his own hometown, being falsely accused, and of course suffering and dying on the cross. At the end of the valleys he was exalted by God and given a name that is above every name.

Apostle Paul had his own share of valleys. As Saul he persecuted Christians (Acts 8:1); then he was converted. Before his conversion Saul was a learned lawyer and trained under Gamaliel, a renowned teacher of the law. (It was like being a clerk of a Supreme Court justice after attending Yale law school.)

On his way to Damascus, Christ met with him and changed his life. As an apostle, Paul experienced many valleys in his life that strengthened him for the rough road ahead in propagating the gospel to the world. Let's consider some of these valleys.

1. Paul was stoned and given up for dead in Lystra (Acts 14:19-20). (The amazing thing to me about this incident was that he went right back into the same city as soon as he was revived.)

2. He broke up with his benefactor, mentor, and friend, Barnabas (Acts 15:36-41).

3. He was imprisoned in Philippi, which eventually led to the jailers' family accepting Christ (Acts 16:16-35).

4. He was arrested, chained, and imprisoned in Jerusalem (Acts 21:27-36).

5. In his letter to the Corinthians Paul wrote about a thorn in his flesh (2 Corinthians 12:7-8) that God would not take away but told him that His grace was sufficient for Paul in that valley.

Paul responded to God by saying, *"I delight in weaknesses, in insults, in hardships, in persecutions, in difficulties. For when I am weak, then I am strong"* (2 Corinthians 12:10).

The story of Job is another good example of God allowing the enemy to unleash hardship as a form of testing in our lives, eventually leading to spiritual and emotional growth. In Job 1:12, we read about God granting permission for Satan to test Job: *"The Lord said to Satan, 'Very well then, everything he has is in your hands, but on the man himself do not lay a finger.'"* Job endured the severe testing (the worst the enemy could lay on him – loss of business, loss of all his children, physical ailments, depression, etc.). After this, *"The Lord blessed the latter part of Job's life more than the first...After this Job lived a hundred and forty years; he saw his children and their children to the fourth generation. And so he died, old and full of years"* (Job 42:12a, 16).

We must not lose sight of God's intent for allowing testing in our lives – the intent is for growth. God allowed the church to be persecuted and then scattered after the death of Stephen, as reported in Acts 8: *"On that day a great persecution broke out against the church at Jerusalem, and all except the apostles were scattered throughout Judea and Samaria"* (Acts 8:1). The result of this scattering was that it brought the gospel to people in Judea and Samaria. A similar process happened in Acts 15, when Barnabas and Paul separated over Mark's coming with them on their journey; this separation again led to the gospel being preached far and wide.

2. God's discipline

God's view of discipline is revealed in Hebrews 12:5-11:

"My son, do not make light of the Lord's discipline, and do not lose heart when he rebukes you, because

the Lord disciplines those he loves, and he punishes everyone he accepts as a son. Endure hardship as discipline; God is treating you as sons. For what son is not disciplined by his father? If you are not disciplined (and everyone undergoes discipline), then you are illegitimate children and not true sons. Moreover, we have all had human fathers who disciplined us and we respected them for it. How much more should we submit to the father of our spirits and live! Our fathers disciplined us for a little while as they thought best; but God disciplines us for our good, that we may share in his holiness. No discipline seems pleasant at the time, but painful. Later on, however, it produces a harvest of righteousness and peace for those who have been trained by it."

The above passage in the Bible clearly indicates the following:

1. God disciplines His children.
2. His disciplines are corrective, not punitive, measures.
3. A lack of discipline indicates that one is illegitimate.
4. When you recognize God's discipline, seek forgiveness.

I have two boys. Ola and I love both of them dearly and we want the best for them. In my own fantasy I have a vision of what both boys will be as adults; in fact, I have said many times that I will believe I'm a success as a father when my boys can say that they know the Lord as their personal savior (and not merely the God of Dad and Mom) and also graduate with professional degrees. To get them to the needed level of spiritual and character development, Ola and I as parents need to set boundaries and be strict. If there is anything I have learned by having two kids, it is that they are different from each other and they require unique ways of dealing with problems. My younger son (Fiyin) is nine years old and my older boy (Ibukun) is thirteen. I have had to discipline the younger more than the older in my years as a parent. Ibukun is a compliant child, while Fiyin pushes the envelope. Each kid requires a different style of discipline. Fiyin is more likely to break a rule, but very quick to say "I'm sorry – it happened by accident." Good luck trying to get Ibukun to say, "Sorry."

God loved the Israelites and chose them as his people (Genesis 12:1-3). Like any father, he disciplines them. Discipline stems from a child's lack of obedience, and obedience is based on a set of rules. As a father who is Holy, Righteous and Just, he gave his kids (the Israelites) a set of rules, which are written in the Books of Leviticus and Deuteronomy.

Let me highlight a couple of the rules that were broken and the discipline that followed.

1. Deuteronomy 5:7 says, *"You shall have no other gods beside me."* This was the first of the Ten Commandments, one that God took seriously.

On his deathbed David charged his son, Solomon, the next King of Israel, with these words: *"So be strong, show yourself a man, and observe what the Lord your God requires: Walk in His ways, and keep his decrees and commands, his laws and requirements, written in the Law of Moses"* (1 Kings 2:2-3). Solomon heeded this advice for the first part of his reign on the throne, but he soon got comfortable and neglected the Law of Moses by **marrying foreign women and worshiping their gods.**

As discipline for disobedience, God pronounced in 1 Kings 11:9-11:

"The Lord became angry with Solomon because his heart had turned away from the Lord, the God

of Israel, who had appeared to him twice. Although he had forbidden Solomon to follow other gods, Solomon did not keep the Lord's command. So the Lord said to Solomon, 'Since this is your attitude and you have not kept my covenant and my decrees, which I commanded you, I will most certainly tear the kingdom away from you and give it to one of your subordinates.'"

The kingdom of Israel was divided in two because of Solomon's sin. His son Rehoboam reigned in Judah, and Jeroboam (Solomon's ex-official) in Israel. As both men divided the land between the two new kingdoms, **Jeroboam built gods (golden calves) in Bethel and Dan**; he also made priests out of those who were not necessarily from the house of Levi, so that his people wouldn't have to travel to Jerusalem to serve Jehovah God. No other Kings of Israel after Jeroboam reversed this abomination. In 2 Kings 17, verse 20 says, *"Therefore the Lord rejected all the people of Israel; he afflicted them and gave them into the hands of plunderers, until he thrust them from his presence."*

Other things could be gods to us, too, not necessarily just images of animals. In Acts 4, while the church was growing by thousands daily, many new Christians sold their possessions just to keep up with the demand and help the church financially. Ananias and Sapphira (husband and wife) sold their possessions, too, but because of their love for money and deceit, gave only a portion of the sale of their property. After they lied and said that they gave everything, they fell and died instantly.

Today many of us have turned things God gave us into gods. Some have made a spouse into a god, others have made a job into a god, turned possessions into gods, or have even made influential or wealthy humans into gods. When we exalt or worship anything or anyone more than God, when that thing takes priority over our relationship and time with God, then we are worshipping it.

2. Another broken rule was that the Israelites were supposed to keep the Sabbath year. Leviticus 25:3-5 says, *"For six years sow your fields, and for six years prune your vineyards and gather their crops. But in the seventh year the land is to have a Sabbath of rest, a Sabbath to the Lord. Do not sow your fields or prune your vineyards."*
Sometime before Joshua led the Israelites into the Promised Land, he told them what they were to do. Concerning the land, he instructed them to farm for six years and let the land rest for the seventh year. This was a command that they never observed while they stayed in the Promised Land for 490 years. In 2 Chronicles 36: 21, the Bible tells us that *"The land enjoyed its Sabbath rests; all the time of its desolation it rested, until the seventy years were completed in fulfillment of the word of the Lord spoken by Jeremiah."*
God himself talked about this discipline in Leviticus 26 and Deuteronomy 28. After giving the law (the Law of Moses), God talked of the blessings and curses. Deuteronomy 28:1-14 is summarized below. If you fully obey (do what you're told immediately and completely) the Lord your God and carefully follow all his commands I give you today, the Lord will:

1. Set you high above all the nations of the earth.
2. You will be blessed in the city and blessed in the country.
3. The fruit of your womb will be blessed; your crops, young livestock and lambs of your flocks will be blessed.
4. Your basket and your kneading trough will be blessed.
5. The Lord will establish you as his holy people.
6. The Lord will grant you abundant prosperity.

But if you don't obey the Lord your God (Deut. 28:15-68):

1. You will be cursed in the city and in the country.
2. Your basket and kneading trough will be cursed.
3. The fruit of your womb will be cursed; your crops, livestock and lambs will be cursed.
4. The Lord will cause you to be defeated by your enemies.
5. The Lord will afflict you with madness, blindness and confusion of mind.
6. You will build a house and not live in it.
7. You will sow and your harvest will be little.

Deuteronomy 28:45 says, *"All these curses will come upon you. They will pursue you and overtake you until you are destroyed, because you did not obey the Lord your God and observe the commands and decrees he gave you."*
To avoid God's discipline, obedience to His laws and decrees are crucial. The wonderful thing about God is that as he disciplines with the left hand, he warmly embraces with the right, letting His children know that what he is doing is for their own good. Hosea 2:13-16 reads:

"I will punish her for the days she burned incense to the Baals; she decked herself with rings and jewelry, and went after her lovers, but me she forgot, declares the Lord. Therefore I am now going to allure her; I will lead her into the desert and speak tenderly to her. There I will give her back her vineyards, and will make the valley of Achor a door of hope. There she will sing as in the days of her youth, as in the day she came up out of Egypt."

3. Sin

Romans 5:12-13 says, *"Therefore, just as sin entered the world through one man, and death through sin, and in this way death came to all men, because all sinned – for before the law was given, sin was in the world. But sin is not taken into account when there is no law."*

1. Sin came into the world as a result of Adam and Eve's Sin – Genesis 3
2. Sin was not taken into account before the law.
3. After the law was given, disobedience to the law made sin significant.
4. Disobedience to the law or sin then resulted in consequences.

The Israelites were given the law of Moses and they were covenanted to obey it (Joshua 24: 1-27), the same way Jesus Christ became the high priest of a new covenant, shown in Hebrews 8:10-11: *"This is the covenant I will make with the house of Israel after that time, declares the Lord. I will put my laws in their minds and write them on their hearts. I will be their God, and they will be my people. No longer will a man teach his neighbor, or a man his brother, saying, 'Know the Lord' because they will all know me, from the least of them to the greatest."*

What persuades us to sin is the enemy (the devil) and the flesh (our will – spurred on by our eyes and mind), and often the two work in tandem.

Let's consider what happened to David in 2 Samuel 11, while his armies were at war. He went on the roof of his house, perhaps because he could not sleep. From his roof he saw a woman bathing in her own house, and he was fascinated by seeing her naked. He used his influence and power to get her (sin number one – adultery). When she got pregnant, he tried covering his adulterous relationship by trying to get Bathsheba's husband, Uriah, sleep with her so it could look like the baby was his (they did not have DNA tests in those days). Uriah refused because he believed having sex with his wife when his soldiers were at war was wrong. David then committed the next sin; he had Uriah murdered by ordering the field commander (through a letter) to place Uriah where the battle was fiercest (so the possibility of his dying at war would be great). Guess what? Uriah died at war.

God sent the Prophet Nathan to David to point out his sin and the consequences, the death of the child born from the affair.

In 2 Samuel 24, we see David's pride got him into a mess (1 John 2:16). He called his commanders and ordered them to count the fighting men in his land. Joab, his trusted commander, tried reasoning with him against taking a census, but David would not hear of it. After the counting, his pride finally crumbled. Verse 10 puts it this way: *"David was conscience stricken after he had counted the fighting men, and he said to the Lord, 'I have sinned greatly in what I have done. Now O Lord, I beg you, take away the guilt of your servant. I have done a very foolish thing.'"*

The consequence for his counting Israel's fighting men was severe; 70,000 men died from a plague sent by God.

King Solomon committed a grave sin by marrying foreign women and serving their gods. In 1 Kings 11:1-4, the Bible says, *"King Solomon, however, loved many foreign women besides Pharaoh's daughter – Moabites, Ammonites, Edomites, Sidionians and Hittites. They were from nations about which the Lord had told the Israelites, 'You must not intermarry with them because they will surely turn your heart after their gods'...As Solomon grew old, his wives turned his heart after other gods, and his heart was not fully devoted to the Lord his God."*

Solomon paid for his sin by God's allowing his subordinate (Jeroboam) to rebel against him and take part of the kingdom away. 1 Kings 11:11 says, *"So the Lord said to Solomon, 'Since this is your attitude and you have not kept my covenant and my decrees, which I commanded you, I will most certainly tear the kingdom away from you and give it to one of your subordinates.'"*

Ananias and Sapphira voluntarily decided to sell their property and give the proceeds to help the fledging church; however, they sinned by lying about the proceeds from the sale of their property, and they both died. While the consequences of their sin seem grave, according to the Bible it brought fear into others who might have been thinking of doing the same thing. Acts 5:11 says, *"Great fear seized the whole church and all who heard about these events."*

Most of us are living a sinful life and doing it knowingly. There is no doubt that under the new covenant brokered by Christ we have grace. 1 John 1:9 says, *"If we confess our sins, he is faithful and just and will forgive us our sins and purify us from all unrighteousness."* In his letter to the Romans, Apostle Paul wrote about the grace of God while we sin, and how we should not continue in sin so that grace might abound.

While we are all sinners and need the grace of God daily, some of us who have confessed Christ continue to live with total disregard for His word and His principles (laws).

Many Christian singles are marrying non-Christians, giving all the excuses in the world (from a lack of men or women to true love), yet in 2 Corinthians 6:14, the Bible clearly states, *"Do not be yoked together with unbelievers, for what do righteousness and wickedness have in common? Or what fellowship can light have with darkness? What harmony is there between Christ and Belial? What does a believer have in common with an unbeliever?"* Many marriages are in turmoil today because of disobedience to this principle, and many marriages with potential are already dissolved because of the union of a believer and unbeliever.

The wages of sin is death. While we have the grace of God and forgiveness, there are consequences for our disobedience.

Chapter Four

The Enemy's Strategy for a Believer in the Valley

The enemy is the Devil, and he has been around a while. He understands that God uses valleys as a teaching tool for a believer. He has read Psalm 23:4: *"Even though I walk through the valley of the shadow of death, I will fear no evil, for you are with me; your rod and your staff comfort me."* He has also read James 1:2-4: *"Consider it pure joy, my brothers, whenever you face trials of many kinds, because you know that the testing of your faith develops perseverance. Perseverance must finish its work so that you may be mature and complete, not lacking anything."*

It is clear, at least to the enemy (not to a lot of believers) that:

1. Testing is for growth.
2. While going through the testing, God is with us.
3. We are supposed to walk through the valley and not dwell in it.
4. Once matured, we will lack nothing in our relationship with Him.

5. Without faith acquired through testing it is impossible to please God.
6. Patience is a necessity in our relationship with others and our walk with God.

Knowing all of these things, the enemy (in my own view) has crafted ways to keep us in a valley instead of walking through it. He uses three main ways to keep us a valley:

1. The enemy hinders our learning and growth.
2. The enemy keeps us in a state of unforgiveness.
3. The enemy keeps us ignorant about our use of spiritual weapons.

1. The enemy hinders our learning and growth.

As indicated above, it's not about how you get into the valley; what is important is what you learn through the valley. Valleys help us grow in faith, learn how to wait on the Lord because His timing is always different from ours, and most importantly mature in our walk with Him.

The enemy uses our pride, fear, anger and self-pity to keep us from learning in the valley.

Pride

Proverbs 16:18 says this about pride: *"Pride goes before destruction, a haughty spirit before a fall."* Pride is thinking too highly of oneself with absolutely no regard for the instructions given by God in His word. Joshua 1:7b says, *"Be careful to obey all the law my servant Moses gave you; do not turn from it to the right or to the left, that you may be successful wherever you go."* Apostle Paul emphasized the importance of humility before obedience in Philippians 2:8: *"And being found in the appearance as a man, he humbled*

The Law of Undulation

himself and became obedient to death – even death on a cross."

Lack of humility or excessive pride hinders our obedience. Let's use the example of a wife who refuses to respect her husband because he does not deserve it (according to her own standard – perhaps he is not making enough money to maintain the lifestyle she wants). Or take a husband who refuses to love his wife because she does not respect him. In marriage we are called to love and respect each other without waiting to see who goes first.

Samson was well aware of the instruction not to marry foreigners. In Judges 14, Samson found a beautiful Palestinian girl named Timnah, and he came home and asked his parents to make arrangements with her family so he could marry her. His parents replied in verse 3, *"Isn't there an acceptable woman among your relatives or among all our people? Must you go to the uncircumcised Philistines to get a wife?"* Feeling invincible because of the strength given to him by God, Samson continued in his pattern of disobedience fueled by pride when he married Delilah (another Philistine). Still with pride of invincibility, he told her the source of his strength, which eventually led to his capture and death.

King David, against the direction of God, counted his fighting men and built up his pride in being in charge of the most powerful kingdom in the world; this brought about the death of 70,000 men in his kingdom.

Jesus told the story of the rich young man in Mathew 19 who came to Christ asking how he could have eternal life. Jesus told him to obey the laws, to which he replied, "I have done that all my life." Jesus then told him to give away his possessions and follow him. Verse 22 says, *"He went away sad, because he had great wealth."* This young man had pride in his possessions, which hindered his obedience to the King of Kings.

Pride puffs up the self, hindering the individual from obeying God's word or following His principles. Pride makes us concentrate on self rather than being selfless. When in a valley, if we are consumed with our pride or the shame brought by the situation, it robs us of the lesson that we could learn from the situation; instead, our emphasis is on what people will say.

Fear

Fear comes from the Greek word *phobos*; it means to terrorize, to demand respect or reverence. The aim of the enemy is to terrorize God's people. When the Bible talks about fearing the Lord it is out of respect or reverence for Him.

America decided after 9/11 that the new war in the world was no longer the cold war but terrorism. Terrorists know they cannot fight and win against the strength of their opponent; they then resort to desperate and cowardly acts meant to intimidate people.

The devil clearly understands that he cannot do anything to a Christian except that which is allowed by God (although most Christians don't know this), so he sets out to tell lies such as "You've committed so much sin, God can't forgive you" or to make Christians tremble at fake powers by demonic activities, like doors closing and opening without a person being there. The enemy also lures Christians into temporary wealth or peace of mind, and then things start to fall apart. In the practice of African religion, the worshippers believe there is a supreme God; however, they are more afraid of the intermediaries like the iron god, the water god, and many more intermediary gods than they are of the supreme God. Thus, they offer daily sacrifices to these gods with fear of retribution if they miss just one day. Bad events are typically associated with missing daily sacrifices. Anything we fear more than God keeps our eyes off of His power and keeps us stuck in the valley.

Iraq is currently going through the process of democracy and at the time of this writing is planning for an election. Local and foreign terrorists are using themselves as human bombs, blowing themselves up in cars and strapping devices on their body. These tactics are putting fear in the Iraqi citizens and some are afraid to join the re-building of their country or participate in the upcoming election.

Fear is meant to paralyze, but when you stand up to fear with the assurance of Christ's love, then you will have victory. 1 John 4:18 says, *"There is no fear in love. But perfect love drives out fear, because fear has to do with punishment. The one who fears is not made perfect in love."*

Anger

Anger in itself is not a sin, but when we entertain it instead of dealing with the underlying issue, then we are giving the enemy a foothold in our lives. Ephesians 4:26-27 says, *"In your anger do not sin, do not let the sun go down while you are still angry, and do not give the devil a foothold."*

Anger is an emotional reaction to a wrong done to us. When we are wronged, we are hurt. A husband whose wife has just committed adultery has every right to be angry. The wife whose husband is on drugs and blows his money on drugs and gambling has every right to be angry. God has every right to be angry when we do things against His character of Holiness or disobey His commandments.

Anger becomes deadly when we don't deal with it or even admit it. The Bible recommends that we deal with anger before the sun sets, something Ola and I have practiced for about eight of our thirteen years of marriage; it has brought peace that flows like a river in our relationship. We still have conflicts, and we know that the other person might be wrong; however, we know that what was done was done with the best intentions or the right motivations. When she corrects my grammar in public (which I hate like most men), I know that she wants the best for me, especially since I speak publicly for a living. I might be angry for a while, but then I learn the lesson and move on. (That doesn't mean I'll not get angry if she does it again.)

The enemy uses anger in many ways to keep us in the valley.

1. Holding on to anger can lead to many physical ailments like depression, high blood pressure, and arthritis. It can also trigger fits of rage, causing an individual to lose control for a brief moment. Many husbands have killed their wives. (Domestic violence is the eighth highest cause of death for women in America and it is the leading cause of injury for women between ages 15-54.) Women are joining the ranks by killing their husbands as well these days.

2. Anger creates an illusion or brings back the past. Husbands or wives that feel wronged could do improper things to shield themselves. They might think of what the spouse did or has done before and either prepare for this battle or calculate how to counter whatever the other person brings, creating an illusion based on perception.

3. The enemy understands that when we hold on to anger we are actually disobeying God and such acts hinder our prayers and communion with God. Mathew 6:12, 14-15 says, *"Forgive us our debts, as we have forgiven our debtors...For if you forgive men when they sin against you, your heavenly Father will also for-*

give you, but if you do not forgive men their sins, your Father will not forgive your sins."

God's desire for us when we are wronged is to get angry, get over it, love the person and then let him (God) avenge if need be. James 1: 19-20 says, *"My dear brothers, take note of this: Everyone should be quick to listen, slow to speak and slow to become angry, for man's anger does not bring about the righteous life that God desires."*

Self-Pity

When a bad thing happens to us, the first question most people ask is, "Why me?" This might be a legitimate question for a while, especially through the process of grieving, but when it starts to get a person down to the point of feeling useless, then the enemy is taking advantage.

As a young man in Nigeria, my family was close to another particular family. Together with their kids we attended Sunday school, choir practice, and many other social functions. The father and mother in this family shared a bond of love that was purely God, as it was different from the other families around us. The man developed gangrene on his left leg and it led to his untimely death. The wife was so devastated that she literarily melted away; she had so much self-pity and did not have the will to stay alive for her kids' sake that she died before the first anniversary of her husband's death. Her death had a big impact on her family, as it destroyed the fabric of the once strong unit. On the contrary, I met a lady who was diagnosed with cancer a couple of months after moving to the United States. She had survived war in her country, and she fought the cancer headlong, not giving up for one minute. Until the day she died, she was still checking out cancer treatment centers that could help her. While fighting, she faced the reality of her terminal illness and made adequate financial provision for her children who would be orphans after her death. This woman's courage in the face of death made her death a celebration.

Self-pity brings attention, and many people enjoy the attention instead of dealing with the situation. It's okay to wonder and question when things happen to us. I ask God, "Why?" when I go through tough times, especially in situations I can't or don't understand. However, when I get what I believe is my answer through reading the Bible, through messages (sermons), during Bible study class, or by sharing with my accountability partners, I swallow my pride and get with God's flow. Job did not curse God like the enemy wanted, but he did wallow in self-pity (like a lot of us), and that did not make God too happy. In Job 38-41, God reminded him who was in charge. God eventually blessed Job beyond what he had before for his obedience and faith through the trials he was put through by the enemy.

2. The enemy keeps us in a state of un-forgiveness.

The enemy tries to keep us in the valley through a lack of forgiveness, both forgiveness of others and of ourselves. In 1 John 1:9, the Bible says, *"If we confess our sins, he is faithful and just and will forgive us our sins and purify us from all unrighteousness."* Romans 12:17-21 says, *"Do not repay anyone evil for evil. Be careful to do what is right in the eyes of everybody. If it is possible, as far as it depends on you, live at peace with everyone. Do not take revenge, my friends, but leave room for God's wrath, for it is written: it is mine to avenge; I will repay, says the Lord. On the contrary: If your enemy is hungry, feed him; if he is thirsty, give him something to drink. In doing this, you will heap burning coals on his head. Do not be overcome by evil, but overcome evil with good."*

The Law of Undulation

After 39 years in this world, all as a Christian, I have come to the conclusion that it takes the grace of God to forgive; it is not something you do because you want to. I spent almost seven years of my life holding on to the wrong I believed my Dad did to the family. When asked, I would always reply, "I have forgiven him," but within me I knew I was still angry and had un-forgiveness. When God finally helped me get rid of my anger, I was free and actually felt better physically. I was able to relax when Dad and I were together. John 15:4 puts it like this: *"Remain in me, and I will remain in you. No branch can bear fruit by itself; it must remain in the vine. Neither can you bear fruit unless you remain in me."*

Forgiveness falls into two categories, forgiveness of oneself for past wrongs and forgiveness of others that wrong you. The enemy uses the lack of both to hold us down in the valley.

The enemy knows that when you don't forgive, the following occurs:

1. Your prayers are hindered (Matt. 5: 23-24, 18: 19)
2. God will not forgive you (Matt. 6: 14-15)
3. God will not judge the offender (Rom. 12: 19-21)
4. You are in bondage (Colossians 3: 15)
5. You are disobedient to God (Isaiah 1: 19-20, 1 John 4: 7-21)

We are all sinners who by the grace of God and the blood shed on the cross of Calvary can receive forgiveness regardless of the sin. Receiving forgiveness does not involve any ritual or rain dance or any funny thing, it only requires a repentant heart, saying, "Lord I'm sorry," "Lord forgive me," and "Lord I will not do it again."This process, however, seems too easy for many people; they feel they need to do something more. Restitution may be done if necessary, but if it will cause more damage to the other person, then don't do it. Some think their sin of abortion, murder, telling lies, or whatever was done before they became a Christian is beyond forgiveness. This is a lie from the father of lies himself. Your being granted forgiveness was not cheap; it took God humbling Himself, taking the form of a man, dying on the cross and paying with His blood to earn you that forgiveness. Titus 3:3-5 says, *"At one time we too were foolish, disobedient, deceived and enslaved by all kinds of passions and pleasures. We lived in malice and envy, being hated and hating one another. But when the kindness and love of God our Savior appeared, he saved us through the washing of rebirth and renewal by the Holy Spirit."*

Learn to forgive yourself. If you don't, you are hindering your prayers, you are calling God a liar (He already said that when you ask He will forgive), and you are carrying an unnecessary burden. On the cross with Jesus Christ in Golgotha were two criminals; one scorned Jesus and the other trusted his life to Jesus and ask for forgiveness after a lifetime of crime. While on the cross Jesus told this repentant "murderer" (death on the cross was for violent crimes) in John 23:43, *"I tell you the truth, today you will be with me in paradise."* Additionally, when you don't forgive yourself and you get into a marriage relationship, your lack of self-esteem or the burden you are carrying will most definitely affect your marriage and may very well destroy it.

Just as we are required to forgive ourselves, we are also commanded to forgive others. My suggestion is to trust God that He can handle the unjust thing done to you. Commit that father or mother who abused you to God. Commit that husband or wife to God. Commit that boss or employer who has wronged you to God. Not only should you let go of the emotional hurt but also learn to pray for those who persecute you.

3. The enemy keeps us ignorant about our use of spiritual weapons.

The enemy tries to keep us in the valley by hindering the use of our spiritual weapons. The weapons of our warfare against the enemy are spiritual, because our enemy is a spirit. The enemy (the devil) realizes that when the weapons are deployed the way they should be, the results are deadly. Philippians 2:10 says, *"That at the name of Jesus every knee should bow, in heaven and on earth and under the earth, and every tongue confess that Jesus Christ is Lord to the Glory of God the Father."*

The weapons of Spiritual warfare are

1. Belt (Truth)
2. Breastplate (Righteousness)
3. Feet/Sandals (Gospel of peace)
4. Shield (Faith)
5. Helmet (Salvation)
6. Prayer (All kinds of prayer)

These weapons are actually a way of life that God expects from a believer, meaning that if we live our lives to glorify God we are insulated from the enemy's attack (except what God allows to mature us and glorify Himself). Consider this scripture, Proverbs 16:7: *"When a man's ways are pleasing to the Lord, He makes even his enemies live at peace with him."*

Knowing all of this about spiritual warfare, the enemy hinders us in two ways:

1. Rendering the weapons useless while we have them. In 1 Samuel 13:16-22, we see a classic example of this. The Philistines were camped beside the Israelites; the enemy instigated them to attack the Israelites like they had done many times before. This time around, the mode of

The Law of Undulation

attack was unique – they sent the Green Berets or Marine Expeditionary Unit into the Israelites' camp and killed all the Israelite blacksmiths. Blacksmiths make weapons. After all the weapon makers were confirmed dead, the Philistines then attacked Israel. The Bible records that there were only two weapons in the whole of Israel – only Saul the King and his son Jonathan had a sword or spear.

The enemy is attacking many of our weapons as we read this book. Some of us cannot tell the truth to save our lives (or we must embellish the truth, so we can look or feel good). Others just can't live the way God wants us to live in Holiness. I know many people who cannot function unless there is a problem with their spouse or family; any time of peace eats away at them. Faith is perhaps the most preached about topic in the Bible, yet many of us don't live in faith. (Until I went through some tough times in my life, faith was merely theory to me.) Living in faith is looking at a situation, doing your best, then standing firm, saying, "Well, it's out of my hands now," not having headaches or depression until the issue is resolved. One of the toughest things to do in a Christian home is maintain daily family prayer; often when it's prayer time, something comes up—the wife says something that offends the husband, or the husband wants to watch football for a little while, then everyone is asleep before he finishes the game.

2. Making us use one weapon instead of another, or making us think that one weapon is better than another. Let's use the analogy of a soldier. He/she has a helmet to cover the head (some with night vision goggles), an armor breastplate to shield the chest region, an M16 rifle and 9mm guns, boots, training in both urban and conventional warfare, and other items like hand grenades. Now let's imagine a soldier going to war equipped with the latest gadgets but without the training; this soldier would not last too long. Or imagine a

soldier with all the equipment except the helmet, or another soldier with everything except the armor breastplate. The fact is that as a believer we cannot choose to pray without a life of holiness, or live a life that is pleasing without a prayer life. As much as we can, we need to utilize every weapon at our disposal, from living a life of faith to living a holy life, from living at peace with people (as much as it is up to us) to praying without ceasing, knowing that our prayers avail much.

We are more effective when all our weapons are deployed together; the enemy knows he stands no chance when this happens. A man or women who lives to please God in both lifestyle and prayer life is a deadly weapon to the camp of the enemy.

4. Making a mountain of a molehill. Another trick of the enemy is to exaggerate our situation so instead of seeking solutions we are looking at this big mountain that cannot be surmounted. The enemy's aim here is for our focus to be on the problem rather than on the person that can fix the problem.

Let's consider some stories from the Bible set in different situations. Ananias and Sapphira had just given their lives to Christ, and about that time believers were selling their property so they could share with other believers who were in need. While this couple had a good heart to give (they actually did sell their properties, after all), they may have panicked about how they were going to exist after they gave everything away. In their fear and lack of faith, they decided to hold some money back after pledging it to God. That led to their death (Acts 5:1-11).

Another example of looking at the situation instead of God is the story of Jesus' commanding Peter to walk on the water. Matthew 14:13-34 paints a picture of what many of us are doing today. The disciples have been around Jesus for a while now, and they've seen many miracles; in fact, they just saw Jesus feed 5000 men with five loaves and two fishes. The disciples went ahead of Jesus in their boat and He was to join them after praying. After praying Jesus decided to join his disciples in the boat, but the boat was quite a distance away from the shore. Jesus then decided to stroll on the water and join His disciples. Because it was dark and they could not see who was walking toward them on the water (they thought it was a ghost), Jesus told them that it was He. Peter then told Christ, "If it's really you, let me walk toward you on the lake as well," to which Jesus replied, "Sure." When Peter concentrated on the command given him by Jesus Christ, he walked on the lake just fine, but when he shifted his focus from Christ to the situation, to the wind that was blowing, he soon began to sink. Jesus had to rescue him. We need to know that whatever the situation is, it is not bigger than our God.

The stoning of Stephen in Acts 7 shows the importance of focusing on God rather than glorifying the situation. Stephen was preaching the gospel of Christ and was arrested by the stooges of the high priest. He was brought before the Sanhedrin (like the Supreme Court of the land, headed by the high priest) to defend himself. Rather than plead his case and beg for his life, Stephen said (in verses 51-53): *"You stiff-necked people, with uncircumcised hearts and ears! You are just like your fathers: you always resist the Holy Spirit! Was there ever a prophet your fathers did not persecute? They even killed those who predicted the coming of the Righteous One. And now you have betrayed and murdered him – you who have received the law that was put into effect through the angels but have not obeyed it."* After his speech, Stephen was stoned to death.

Looking at Jesus, the author and finisher of our faith, is what is required of us when in tough situations. While this is not easy (trust me, I know), we should look at the joy set before us on the other side of the situation. When we choose to look at the magnitude of the problem (which could be severe and unbearable), then we start to sink like Peter, which is exactly the plan of the enemy for us in the valley.

5. The Isolation rule – the enemy tries to isolate us. The Bible is very clear about having a support system. At creation in the book of Genesis, after God created man, he saw that man was alone (which was not good). He made man fall into a deep sleep and made a woman from his rib.

Reading through the Bible, we also see that God-the-Father works in tandem with God-the-Son and God-the-Holy-Spirit. Jesus said that He only does the will of His Father and the Bible says the Holy Spirit glorifies the Son.

King Solomon in Proverbs 15:22 said, *"Plans fail for lack of counsel, but with many advisers they succeed."* Jesus, understanding the importance of not being alone, said that where two or three shall gather (assemble) in His name, there He will be also. He told his disciples to gather together in Jerusalem until the visitation of the Holy Spirit that would empower them. When the apostles set out on their various journeys, they rarely traveled alone. Barnabas and Paul traveled together, and after they separated, Barnabas took Mark with him and Paul took Silas and Timothy with him.

The enemy understands that when two or three gather in God's name, it is a threat to his kingdom. His game plan, then, is to isolate us for the following reasons:

- When alone, our minds go crazy with thoughts that sometimes do not edify us and glorify God. The number one reason that people in ministry fail, according to statistics, is lack of accountability.
- When we are able to share our burden with someone else (therapy), even if the person says nothing back to us, it clears our mind for that time period.

- Coming together with other believers twice or thrice a week encourages us. As a lay counselor, I know that people get satisfaction from knowing that other people within the body are going through the same struggles they are going through.
- When you share your ideas with someone else, they most likely have a different concept of the issue. In seventeen years of marriage, Ola and I have often held different views on issues. To me that is exciting because I know the issues were dealt with both emotionally and analytically.

The enemy uses pride and shame to isolate us.

Some parents are so concerned about how the family is viewed by the public that when one of their kids does something wrong, their action is mitigated by what they believe people will think rather than the best interest of the child. An example is a family whose teenage daughter gets pregnant; because of the shame, they ask their daughter to abort the child instead of facing the shame and turning the situation into a positive one.

When we find ourselves in the valley, let's do our best to get rid of the shame and pride. After all, Christ left his glory in heaven to come into this world and be born in a shameful place, yet His endurance and the determination to do the will of the Father kept Him going.

Chapter Five

What To Do in the Valley

King David, perhaps more than many people, went through many valley situations in His life – running from Saul, running from his own son who wanted the throne, enduring discipline from God after he counted his fighting men, the effects of his sin of adultery with Bathsheba – and through all these valley experiences he understood that valleys make us strong. He wrote in Psalms 23:4 that *"Even though I walk through the valley of the shadow of death, I will fear no evil."* He understood that a valley is meant to be walked through; we are not supposed to lounge in it.

When we find ourselves in a valley situation for whatever reason (because God allows the enemy to inflict some wounds, or He chooses to discipline us as a father does his children, or we put ourselves in such situations through sin and disobedience), we need to realize that there is already a way out for us. Christ died on the cross more than 2000 years ago and everything we need, including forgiveness and power over the enemy, was completed when He said, "It is finished."

When in a valley, what you should do is:

1. Recognize the enemy's game plan and use our spiritual weapons to fight this spiritual enemy (1 Peter 5: 6-11, Ephesians 6: 10-18)

2. Know and seek the grace of God (1 John 1:9, Psalm 51, 2 Samuel 24:14)

3. Humble yourself and be obedient (2 Samuel 12)

1. Recognize the enemy's game plan and use spiritual weapons

The devil is an enemy of God. He wants as many people as possible to spend eternity in hell with him, and while on earth, he'll make life as miserable as possible using whatever tricks possible. The enemy's aim is summarized in 1 Peter 5: 8: *"Your enemy the devil prowls around like a roaring lion looking for someone to devour."*

Let's examine the above scripture carefully. The enemy prowls around – this brings to mind images of a pedophile or kidnapper scouring the internet or sneaking around neighborhoods looking for kids that are vulnerable and unprotected to pounce on. Many of these prowlers are successful with kids whose parents don't monitor their internet usage or who leave the kids alone for too long by themselves. The enemy does the same; he is going back and forth on earth looking for non-Christians and Christians who will be disobedient. The enemy was successful with Eve in Genesis 3, but unsuccessful with tempting Christ after He fasted for forty days and nights. In the book of Job, God allowed the devil to tempt Job; the enemy could not do anything to Job until God said it was okay. The enemy can also appear as an angel of light to deceive Christians. In the above passage, he is like a roaring lion, which means he is not a lion but roars like one. There is only one lion and that is the Lion in the tribe of Judah (Jesus the Christ). When the enemy roars like a lion, his intention is to instill fear into people (1 Samuel 17:1-11). This could be the fear of change (Matthew 19:22), the fear of the unknown (Acts 5:1-11), the fear of present circumstances (Numbers 14:1-4), or fear instilled through physical suffering (Job 2:4-8).

The activities of the enemy occur in the spirit realm (the heavenly realm in Ephesians 1:3) and the enemy himself is a spirit (Job 1:6-7). In Ephesians 6:10-12, Apostle Paul describes our spiritual battle this way: *"Finally, be strong in the Lord and in His mighty power. Put on the full armor of God so that you can take your stand against the devil's schemes. For our struggle is not against flesh and blood, but against the rulers, against the authorities, against the powers of this dark world and against the spiritual forces of evil in the heavenly realms."* To fight spirits in the heavenly realms as human beings, Christ has given us spiritual weapons, listed in verses 13-18:

> *"Therefore put on the full armor of God, so that when the day of evil comes, you may be able to stand your ground, and after you have done everything, to stand. Stand firm then, with the **belt of truth** buckled around your waist, with the **breastplate of righteousness** in place, and with your **feet fitted with the readiness that comes from the gospel of peace**. In addition to all this, take up the **shield of faith**, with which you can extinguish all the flaming arrows of the evil one. Take the **helmet of salvation** and the **sword of the Spirit, which is the word of God**. And **pray in the Spirit** on all occasions with all kinds of prayers and requests."*

Let me reiterate our Spiritual weapons and delve more deeply into their application:

1. Belt of **Truth**
2. Breastplate of **Righteousness**
3. Sandals of **Peace**
4. Shield of **Faith**
5. Helmet of **Salvation**
6. Sword, which is the **Word**
7. **Prayer** – All kinds of prayer and in the Spirit

Our spiritual weapons are actually a lifestyle that pleases God, living in truth and speaking the truth in love, even when it hurts. Many times in our lives we exaggerate a topic to make ourselves look better (I've been guilty of that); it's called stretching the truth. Sometimes we lie just to cover something we've done, which most of the time results in many lies to cover the first one. Telling the truth is like a belt; it holds up the other stuff we put on. Living a righteous life protects the vital organs in a Christian life like the breastplate protects a soldier. Proverbs 16:7 says, *"When a man's ways are pleasing to the Lord, he makes even his enemies live at peace with him,"* meaning that those who plan to do you harm will bless you instead (Read Numbers 22-24, the story of Balaam trying to curse the Israelites). A Holy life brings us directly to God and extends our relationship with him in his presence. Psalms 15:1-2 says, *"Lord, who may dwell in your sanctuary? Who may live on your holy hill? He whose walk is blameless and who does what is righteous, who speaks the truth from his heart."*

Living at peace is a requirement from God. Romans 12:17-18 says, *"Do not repay anyone evil for evil. Be careful to do what is right in the eyes of everybody. If it possible, as far as it depends on you, live at peace with everyone."* This passage indicates that living at peace with people may be impossible in this wicked world, but from our own side of the story, we should do our best to live in peace, not seeking revenge. Jesus took this further by saying that we should love those who persecute us.

Another thing I've discovered in my life is that God's peace does not mean the absence of trials or problems; it means calm in the midst of the storm. It means having the assurance that God is your rear guard even when everything around you is falling apart. When finances become tight (even when you have been faithful), when your spouse cheats on you (when there is no reason to do so), when your church members rebel against you (purely because of your stand for God in a relative truth world), or when odd things happen to you (like when I was not allowed entry into Manchester, England for a conference because of a bogus reason after many visits to the same country for the same purpose), we need to realize that His Grace is sufficient for us.

The Bible says that without faith it is impossible to please God. We need to believe that he will do that which he has said. When faced with hard situations, God wants us to trust him. Abraham trusted God when he was told to sacrifice his promised son on Mount Moriah. While it did not make sense to him, he still trusted God. Abraham trusted and had faith in God when told that he would have a child when he was close to a hundred years old and when his wife was way past menopause. By faith, Joshua (a four star general tested in many battles), dropped his weapons and marched around Jericho as instructed, defeating a fortified city without losing a soldier. By faith the Apostle Paul went from one city to another proclaiming the good news of Christ, knowing the suffering and persecution that awaited him at every turn. By faith, believers gathered together on Pentecost as instructed by Christ to await the Holy Spirit that would empower them.

While speaking to the Sanhedrin, Peter made this bold statement as recorded in Acts 4:12: *"Salvation is found in no one else, for there is no other name under heaven given to men by which we must be saved."* Our salvation lies in our believing the Lord Jesus Christ. This belief brings us under a new covenant with God and earns us the right and authority to use the name of Jesus. Philippians 2:9-11 says, *"Therefore God exalted him to the highest place and gave him the name that is above every name, that at the name of Jesus every knee should bow, in heaven and on earth and under the earth, and every tongue confess that Jesus Christ is Lord, to the glory of God the Father."*

One of the most potent weapons we have is the word of God, first in the person of Christ (John 1:1 says, *"In the beginning was the word, and the word was with God, and the word was God."*) and second in terms of the written word (Hebrews 4:12 says, *"For the word of God is living and active, sharper than any double-edge sword, it penetrates even to dividing soul and spirit, joints and marrow; it judges the thoughts and attitudes of the heart."*) In his letter to Timothy, Apostle Paul wrote these words about the written word in 2 Timothy 3:16: *"All scripture is God-breathed and it is useful for teaching, rebuking, correcting and training in righteousness, so that the man of God may be thoroughly equipped for every good work."* The word that became flesh now has a name that we can call upon to defeat our enemy; the written word is our instructional guide in how to use our spiritual weapons.

As believers we need to be aware that using our weapons is not a choice because the warfare (spiritual) we are in is not our choice either. These weapons should be deployed when needed and always ready. We should not wait for a valley experience to pray, or to live a righteous life, or to read the word of God and apply it to our lives. This should be our way of life. Reading the Bible daily is a habit we must form and it must start with our children. Our older boy, Ibukun, is a Kanakuk Kamper, and they are encouraged to do daily

Bible readings called "devos" (from devotion), with rewards such as having their name on a walk of fame in camp. He also attended BSF (Bible Study Fellowship) with me, completing seven years. Just like Ibukun, I too felt forced into all this Bible stuff when I was his age. I reassure him that one day as a Dad he'll thank me for forcing him, just as I now appreciate my parents making us go to church, which has served as a solid foundation in all of our lives (my siblings and myself).

Our weapons must also be fitted properly. Imagine one of our soldiers in the various theatres of war worldwide (especially in Iraq and Afghanistan), going to fight without his helmet on, or going to fight without his bulletproof vest, his shoes, or better still without his gun. In order to go into battle and fight effectively, a soldier puts on all his gear, from the boots to the helmet. Without a piece of his gear, his superior would perhaps not let him go into the war zone.

Most of us Christians are fighting our spiritual warfare with incomplete gear. Some think, "As long as I pray, living a holy life is something for pastors only." For others, living at peace with people around us seems like an insurmountable challenge (and some family members cannot operate without fighting someone), yet we are commanded to be at peace. Let's make praying a habit also, setting aside a time that is convenient, night or day, as well as praying throughout the day – in our car or even while in the bathroom. Prayer allows us to fellowship with our Creator. Don't just talk to him; listen to him as well. He really wants to talk to you.

2. Know and seek the grace of God

It is important for a Christian to know and be consciously aware that there is already a solution to any problem he/she might encounter in this world. In John 16:33, Jesus said the following to his disciples (and to us): *"I have told you these things, so that in me you may have peace. In this world you will have trouble. But take heart! I have overcome the world."*

The death of Christ on the cross at Calvary is a definitive victory over whatever problem we have. With his last breath Christ pronounced to the heavens and earth, "It is finished," meaning there is no other "work" that needs to be done by you or me to attain victory in our struggle; all we need to do is surrender to him and be obedient to his words. Jesus himself explained this better in John 15:1-4 when he said, *"I am the true vine, and my father is the gardener. He cuts off every branch in me that bears no fruit, while every branch that bears fruit he prunes so that it will be even more fruitful. You are already clean because of the word I have spoken to you. Remain in me, and I will remain in you. No branch can bear fruit by itself; it must remain in the vine. Neither can you bear fruit unless you remain in me."* Let's highlight some important points Jesus said in the scripture above:

1. Branches without fruit will be cut off, which means we need to bear fruit.
2. Branches with fruit will be pruned for better growth.
3. To bear fruit, a branch must remain in its source of life, which is the vine.
4. Anyone attached to the vine is clean, because of the nutrients (words) coming from the vine.

According to Galatians 5, fruit is Godly characteristics like forgiveness, joy, peace, patience, gentleness, self-control, kindness, faithfulness, love, and many more. In other words, the vine has done all the work and the branch just needs to get plugged in and enjoy the ride. The vine was once a seed sown in the ground that died and then resurrected, grew deep roots, and finally became a huge vine (tree) that feeds many branches.

The Law of Undulation

With this understanding, we realize that as Christians, all we need to do is trust Christ and be obedient to His words because He has already finished the work.

When He says in 1 John 1:9 that *"If we confess our sins, he is faithful and just and will forgive us our sins and purify us from all unrighteousness,"* then we should believe it, instead of believing the enemy's lie that our sin is so bad we can't be forgiven. While a clergy volunteer in prison ministries for about fourteen years, I met many inmates who, after accepting Christ, still thought that just confessing was not enough for all the bad things they had done. I always reminded them that Christ already paid the price and that He wanted them to live a lifestyle that glorifies Him in return.

It is typical of the enemy to bring up one's past—the abortion, the lying, the murder, the drug and alcohol abuse, the sexual immorality, and many more things that we are ashamed of. It is more important for us as believers to trust that once we confess (repent of) our sins, he will forgive. If we believe that our sins are beyond forgiveness, we become depressed and basically useless in other areas of our lives, or we think that by doing good work we'll make up for the bad things we have done.

The grace of God is available for anyone what asks for it. This grace is not the absence of problems or difficult times. In 2 Corinthians 12:9, Paul asks God to take away his pain (what sort of pain is not revealed in the scripture). According to him, God told him that His grace (God's) is sufficient for him (Paul). King David knew God had forgiven him after he counted the fighting men of Israel when he should not have. He also knew that he had to pay for the sin. When presented with choices of consequences, he chose one and made what I think is the best statement by anyone who understands the grace of God in 2 Samuel 24:14: *"I am in deep distress. Let us fall into the hands of the Lord, for His mercy is great; but do not let me fall into the hands of men."*

In whatever situation or circumstances that you find yourself, always remember that God's grace is sufficient for you.

3. Humble yourself and be obedient

Lack of obedience is perhaps the most important weapon of the enemy to keep a believer in the valley. Let's look at an example.

John and Mary had been married for fifteen years. John was a laid back person who took life too easily, per to his wife. "He is not a self-motivator," she'd always say. Mary, on the other hand, believed you don't get anything for free; you had to go get it. She was successful as a small business owner. The years of her carrying the family financial burden was starting to take its toll on her mind; she thought the marriage was actually keeping her from the place she had dreamt she'd be at by this time of her life.

Many families reading this book are like John and Mary, a self-motivated wife and a laid back husband. According to the Bible, that laid back husband is still the head of the household. What the self-motivated wife can do is pray for her husband while bringing up issues and suggestions to him (not mandating anything). There is also a place for tough love if the man is just plain lazy. To enjoy her marriage, Mary needs to be obedient to the word of God. Nagging him to change will not solve anything; rather, it will kill his self-esteem. Showing love to him, lovingly bringing up issues, and praying for the situation will go farther. This approach may sound ridiculous and unfair, but it is a Biblical principle.

To obey, we need to first humble ourselves. The Bible says that Christ humbled Himself and became obedient unto death.

Humility

Humility in our world today is often seen as not sharing or bragging about your possessions or a great attribute you've been blessed with. For instance, it is considered pride for me to say, "I've written this many books or traveled to this many countries." While bragging could be a sin (depending on the motives and circumstance), it is not always a sin. In Numbers 12, Moses (the author of the book of Numbers) called himself "the most humble man in the world."

True humility is a total dependence on God (Exodus 33:12-16, John 15:1-4), the way you see yourself (Philippians 2:6-7, Luke 14:8-11) and the determination to obey God's word (Joshua 23:6-11).

Total dependence is when you realize that everything in your life, your achievements, your family, your position is absolutely based on the mercies of God, and without God, you can achieve nothing. Moses realized this when he said he would not lead the Israelites one step without God leading the way.

Many of us often see ourselves based on our position (whatever position we deem prestigious). One might say, "I'm a senior pastor or a Bishop, so I should have this many perks and this big of an entourage when I travel." Another may say, "I'm a CEO, so I'm invincible," or "I'm the president of a country, so my powers are unlimited." The problem with seeing ourselves from a positional standpoint rather than as an instrument of God is that flesh and pride take over and we behave like a small god. It is common these days to have pastors with security entourage and many assistants.

I know many senior pastors who, because of the comfort of the current position and the fear of unknown, will not leave their position to attend a higher calling from God. (They know it in their spirit.) Many CEOs are so comfortable in their positions that they do improper things because they think they are invincible. One of the hallmarks of the start of the 21st century is the corruption of CEOs and the focus on personal enrichment to the detriment of quality and care for their staff.

There is a saying in Washington DC that someone "drank out of the Potomac river" when a good person from rural America comes in as a congressman or senator and then becomes corrupt just to stay in power. Many politicians who have fallen from grace say that after many years of having others make you feel like a god, you start to believe it and soon engage in corrupt and immoral practices, thinking you can get away with almost anything.

Obedience

In 1 Samuel 15, King Saul had just dug himself into a spiritual valley through his disobedience (which is sin). God had told him to kill everything when fighting the Amalekites; Saul instead kept the best animals and Agag, the king of the Amalekites. When confronted by Prophet Samuel about what he did, instead of apologizing and repenting of his sin, Saul again lied by saying that he did obey God (to a large percentage) and that the animals they brought back were for sacrifices to God (basically lying and giving excuses). Samuel told Saul that God had departed from him, as he was still not repentant. Saul's biggest concern was that the Prophet would appear with him in front of the people so they wouldn't know that God had departed from him.

Let's contrast this story to that of King David in 2 Samuel 11. After sleeping with Bathsheeba, who ended up pregnant, then trying to get her husband to sleep with her, and finally arranging for her husband Uriah to be killed, David was certainly in the wrong. God sent Prophet Nathan to confront David about what he did. Instead of giving excuses like Saul, David regretted his deeds and asked God for forgiveness, mourning for several days.

Let's make this personal. In the valley you are in as you read this book, are you giving excuses for wanting to do things your way, or are you surrendering to do what God wants?

Obedience is doing what you're told immediately and completely. King Saul figured he had at least done about 95 percent of what God wanted, sparing only King Agag and the fat animals that could be used for sacrifices. While his reasoning made human sense, it was not what the instruction required. Like Prophet Samuel told Saul, "Obedience is better than sacrifice." Is God asking you to honor your husband or love your wife, but you are giving excuses about why she is unlovable or why you can't honor him? Or are you like David, who realizes that repenting of sin and then doing what God wants will get you out of the valley?

David had two chances to kill Saul when Saul wanted to kill him (1 Samuel 24:4, 26:8-14), but on both occasions, he was obedient to the word of God that says, "Touch not my anointed and do my prophets no harm." After David spared his life twice, Saul kept seeking to kill David. Are you persistently being obedient even when you are being paid evil for your good?

The enemy wants you to remember that your Saul is not a good thing. Even when you spare his life, he'll still come after you. God wants you to remember that obedience will eventually reap a reward (Galatians 6:9).

Many people are in emotional valleys because of unforgiveness. The Bible is very clear about forgiveness. Matthew 6:12 says, *"Forgive us our debts, as we also forgive our debtors,"* and verses 14-15 add, *"For if you forgive men when they sin against you, your heavenly Father will also forgive you, but if you do not forgive men their sins, your heavenly Father will not forgive you."* Colossians 3:13 says, *"Bear with each other and forgive whatever grievances you may have against one another. Forgive as the Lord forgave you."* Jesus even told a parable about forgiveness (Matthew 18:21-35).

We are commanded to forgive, and the lack of forgiveness hinders our prayer and communion with God. Paul specified in his letter to the Colossians to forgive *"whatever grievances"* (Colossians 3:13). In counseling I hear husbands and wives give excuses not to forgive, everything from adultery to an abusive ex to absentee parents to a mean boss and many more. When we don't forgive, we bear a grudge against the other person; such a grudge can become detrimental to our own health, apart from hindering us spiritually. We can become dependent on medication for anxiety, depression, stress, and other mood-related diseases.

Are you in a financial valley as you read this book? Then purpose to follow what the Bible says. In Luke 6:38 Jesus said, *"Give, and it will be given to you. A good measure, pressed down, shaken together and running over, will be poured into your lap. For with the measure you use, it will be measured to you."* According to Jesus, the main key to being blessed financially is giving even when you don't think you have enough to give.

Let's consider the case of Elijah and the Widow of Zeraphath (1 Kings 17:7-24). Elijah was the Prophet of God fighting against the idolatry of the government of Ahab and his wife Jezebel who had prostituted the worship of God for Baal. Elijah pronounced God's judgment of famine on Israel. While the brook that fed him dried up, God sent him to a widow in Zeraphath. When Elijah got there, the widow had just enough food to last a short while. 1 Kings 17:12 says, *"'As surely as the Lord your God lives,' she replied, 'I don't have any bread – only a handful of flour in a jar and a little oil in a jug. I am gathering a few sticks to take home and make a meal for myself and my son, that we may eat it – and die.'"* In verse 13 Elijah convinced the widow to still make some bread for him with what she had left. She had a choice to believe what God said through Prophet Elijah or to go by what she had physically. In verse 15 we read that she chose to believe the prophet and according to the Bible, the flour and oil were never used up in her house. In 2 Corinthians 9:6, Apostle Paul reminds us that *"Whoever sows sparingly will also reap sparingly, and whoever sows generously will also reap generously. Each man should give what he has decided in his heart to give, not reluctantly or under compulsion, for God loves a cheerful giver."*

As believers, we should not let the enemy deceive us about our giving. Such deceits include:

1. "I don't have enough money for my bills, so why should I give to the church?" As a lay counselor who has been privileged to counsel many married folks, when we do a budget (planning income against expenditure), most of the time the income is sufficient but habits need to be changed with the help of the Holy Spirit. Like the widow of Zeraphath, believe the word of God; the Bible says God wants us to try him in the area of our giving (Malachi 3:10).

2. "The pastor is living a luxury life." Pastors are like any other human beings; they can have insatiable tastes as well. Today, we have pastors who openly flaunt wealth and even challenge people. To such, I would reference the words of Apostle Paul in 1 Corinthians 9:1-27 and 1 Corinthians 10:23-24. On the other hand, believers should realize that pastors should be adequately compensated. Most of us would pay our lawyers $350 an hour if needed. As a society we pay pro-players ridiculous amount of money, yet we question the salary of our pastor who is being used by God to take care of us both spiritually and emotionally. We also need to understand that our giving

is not to the pastor but to God. While God does not spend human currency, He uses humans to do his will in the world (see Luke 6:38).

Some of us are in valleys because of our location (geographical valley). In 2000, when my family left New Jersey for Omaha, a lot of people were skeptical about the move. While I was very sure of the move myself, I sometimes wondered if I had heard God correctly because I resigned a good job and landed in Omaha on July 1, 2000 without a job. Today, Omaha has been a blessing to my family. I'm now in a full time ministry that reaches worldwide, my wife has her own clinic (these were not reasons why we came to Omaha, however), our kids love their school and we all love the church and are very active in various departments. And these are things we consider important.

Abraham is another person who had to move to receive God's blessing. In Genesis 12:1, God tells Abram to leave his country, his people, and his father's household, and when he gets to the land that God will show him, He (God) will make him a great nation and make his name great, a promise that was fulfilled after his death and through his descendants. Elijah was another person who had to move. After pronouncing to King Ahab that there would be famine in Israel, Elijah was directed to go to Kerith Ravine, East of Jordan, where God fed him through ravens (birds) and gave him water from a brook. When it was time to leave, God allowed the brook to dry up and told Elijah to go to Zeraphath, where a widow would feed him. In the book of Acts, we read many times of God directing Apostle Paul where to go and where not to go (Acts 16:1-10).

Let me ask you these questions: Where is God asking you to move? Are you sure it's God? Why are you holding back? Is the brook dry where you are now? Is it that you just can't bear to leave the comfort of your present abode and the enemy is encouraging you to stay on? Let me assure you that if God has asked you to move, He has already made provision for you at the new location. Just be obedient.

As believers we should always remember that, however we get into the valley, God can and will turn things around for us if we surrender to His will, and what the enemy meant for evil will become an opportunity for us to grow in faith. It will help us develop perseverance and mature both spiritually and emotionally.

Chapter Six

God's View of Your Valley

As Moses wrapped up his time of leading the Israelites, God assured him that He would never leave nor forsake him (Deuteronomy 31:6). The writer of Hebrews echoed this same sentiment of God to the church in Hebrews 13:5b: *"Never will I leave you; never will I forsake you."*

A great insight into God's view of a valley is prophet Jeremiah's letter to the remnant of Israel during the exile in Babylon. *"This is what the Lord says: 'When seventy years are completed for Babylon, I will come to you and fulfill my gracious promise to bring you back to this place. For I know the plans I have for you', declares the Lord, 'plans to prosper you and not to harm you, plans to give you hope and a future.'"* (Jeremiah 29:10-11)

Prophet Jeremiah's letter came to a group of people who were residing in a foreign land, a land to which their forefathers were taken into captivity. Seventy years later, they were sort of free, but like a dog, they had freedom but didn't even know it. While they were in this predicament, God came in with a great comforting line that He's got a great plan for them, which included great prosperity (not just financial wealth).

Let's do a quick overview of some men and women of God in the faith hall of fame. Jonah was given an assignment by God, to go and warn the people of Nineveh of the impending judgment of God. Jonah did not salute and say, "Yes Sir!" We read that he actually ran away, traveling on the opposite side of Nineveh, thinking God could not get him. After three days in the belly of a fish (a valley experience), he reluctantly changed his mind and went to Nineveh.

Jacob, the deceiver, is actually in this hall of fame. This is a guy who made his brother give up his birthright with a bowl of porridge, tricked his father and got blessed instead of his brother, and tricked his father in-law, Laban, after they separated their flocks. Somewhere along the line he changed and dedicated his life to God, and then his name was changed to Israel.

Abraham and Sarah are also interesting people in the hall of fame. Choosing Haggai for her husband was definitely not God's plan. Abraham's moving his family to Egypt because of famine and subsequently lying to Pharaoh that Sarah was his sister is not an attribute of a faith hall-of-famer.

The lesson for all of us is that God uses us in spite of us and He allows us to make a lot of mistakes. Sometimes He even allows the devil to touch or harass us (see the story of Job). Each one of the people learned valuable lessons in their various valley times that were instrumental to their walk with God and their mention in the faith hall of fame.

To God, your valley (however you get in it – sin, discipline or disobedience) is not a big deal because He can and does make everything work together for the good of those who love Him and are called according to His purpose (Romans 8:28). His intent, like Apostle James said, is that you grow in faith, you grow in patience and reliance on him, and most importantly that you become mature so He can use you to impact others. In John 15:8, Jesus suggests that the reason God wants us to bear fruit is not really for us (a tree never takes of its own fruit!), but because our fruit brings glory to the Father and shows that we are His disciples.

God used forty years in the wilderness of Midian to teach Moses patience and total reliance on God rather than on personal ability. God used the prison to bring Joseph in contact with his destiny and another two years to solidify his maturity (displayed in his act of forgiveness). God used the death of Stephen to get the comfortable Christians out of Jerusalem – sharing the gospel to the ends of the world as they ran from the Pharisees and other members of the Sanhedrin council.

I believe that God is more concerned with our relationship with Him and in us as a finished product that brings forth good fruit. Whatever route we take to be what He desires as a finished product is totally inconsequential to Him. Being our Creator, He knows we have a very short memory as humans and that our personalities are different, so intermittent valley experiences after a great mountaintop period jolts us back into reality.

As an itinerant minister, I've come to realize that my prayer life is more sharpened and focused when I have down time (or deliberately create it within a busy schedule) and that it nose dives when I have a full calendar, speaking almost on a weekly basis when my emphasis is on getting ready for each speaking engagement. As much as He is glorified in my speaking and writings, His biggest desire for Femi and each of us is a personal intimate relationship on a daily basis, whatever that looks like for each person.

The purpose of Joseph was to save his people from famine (God told Abraham that his descendants were going to be in captivity, way before Isaac and Jacob were born); the slavery and prison events were just character builders along the way. The purpose of David was to lead his people; working with a 600 member rag-tag army, running from one cave to another, seeking refuge with the Philistines after he had just killed their champion Goliath, were just character builders for him to succeed as a king both at war and in diplomacy.

The questions we all need to ask ourselves (instead of complaining about the valleys we find ourselves in) are, "How did I get here? What is God's purpose? What character traits am I supposed to learn from this valley experience?" And then ask God for the grace and strength He promised us as we walk through the valley.

REFERENCES

Goodstein, Ellen. "Unlucky lottery winners who lost their money." Bankrate.com. (March 29, 2006). http://www.bankrate.com/brm/news/advice/20041108a1.asp.

Ikomi, Tai. "His Beauty for My Ashes." New York: Triumph Publishing, 1989.

Lewis, C. S. "The Screwtape Letters." New York: Harper Collins, 2001.

Moore, Gordon. "Going Solo: Making the Leap." American Academy of Family Physicians. (February 2002). http://www.aafp.org/fpm/2002/0200/p29.html.

Zondervan NIV Bible. Grand Rapids, MI: Zondervan, 2006.

www.ingramcontent.com/pod-product-compliance
Lightning Source LLC
Chambersburg PA
CBHW030915180526
45163CB00004B/1847